NUREG-1855
Vol. 1

U.S.NRC
United States Nuclear Regulatory Commission

Protecting People and the Environment

Guidance on the Treatment of Uncertainties Associated with PRAs in Risk-Informed Decision Making

Main Report

Manuscript Completed: March 2009
Date Published: March 2009

Prepared by:
M. Drouin[1], G. Parry[2]
J. Lehner[3], G. Martinez-Guridi[3]
J. LaChance[4], T. Wheeler[4]

[3]Brookhaven National Laboratory
Upton, NY 11973

[4]Sandia National Laboratories
Albuquerque, NM 87185

M. Drouin, NRC Project Manager

[1]Office of Nuclear Regulatory Research
[2]Office of Nuclear Reactor Regulation

ABSTRACT

This document provides guidance on how to treat uncertainties associated with probabilistic risk assessment (PRA) in risk-informed decisionmaking. The objectives of this guidance include fostering an understanding of the uncertainties associated with PRA and their impact on the results of PRA and providing a pragmatic approach to addressing these uncertainties in the context of the decisionmaking.

In implementing risk-informed decisionmaking, the U.S. Nuclear Regulatory Commission expects that appropriate consideration of uncertainty will be given in the analyses used to support the decision and in the interpretation of the findings of those analyses. To meet the objective of this document, it is necessary to understand the role that PRA results play in the context of the decision process. To define this context, this document provides an overview of the risk-informed decisionmaking process itself.

With the context defined, this document describes the characteristics of a risk model and, in particular, a PRA. This description includes recognition that a PRA, being a probabilistic model, characterizes aleatory uncertainty that results from randomness associated with the events of the model. Because the focus of this document is epistemic uncertainty (i.e., uncertainties in the formulation of the PRA model), it provides guidance on identifying and describing the different types of sources of epistemic uncertainty and the different ways that they are treated. The different types of epistemic uncertainty are parameter, model, and completeness uncertainties.

The final part of the guidance addresses the uncertainty in PRA results in the context of risk-informed decisionmaking and, in particular, the interpretation of the results of the uncertainty analysis when comparing PRA results with the acceptance criteria established for a specified application. In addition, guidance is provided for addressing completeness uncertainty in risk-informed decisionmaking. Such consideration includes using a program of monitoring, feedback, and corrective action.

Paperwork Reduction Act Statement

Public Protection Notification

FOREWORD

In its safety philosophy, the U.S. Nuclear Regulatory Commission (NRC) always has recognized the importance of addressing uncertainties as an integral part of its decisionmaking. With the increased use of probabilistic risk assessment (PRA) in its risk-informed decisionmaking, NRC needs to consider the uncertainties associated with PRA. Moreover, to ensure that informed decisions are made, NRC must understand the potential impact of these uncertainties on the comparison of PRA results with acceptance guidelines. When dealing with completeness uncertainties, NRC also must understand the use of bounding analyses to address potential risk contributors not included in the PRA. Ultimately, when addressing uncertainties in risk-informed decisionmaking, NRC must develop guidance for those uncertainties associated with the PRA and those associated with our state of knowledge regarding design, etc., and the different decisionmaking processes (e.g., expert panel). In addition, this guidance should cover the various stages of the plant that the PRA is assessing (i.e., design, construction, and operation); the different types of reactors (e.g., light water reactors [LWRs] and non-LWRs); the different risk metrics (e.g., core damage frequency, radionuclide release frequency); and the different plant-operating states and hazard groups.

At this time, the focus of the guidance developed is for addressing uncertainties associated with PRAs. Although the process discussed in this document is more generally applicable, the detailed guidance on sources of uncertainty is focused on those sources associated with PRAs assessing core-damage frequency and large early-release frequency for operating at-power LWRs considering internal events (excluding internal fire). Work beyond this scope has been initiated, and NRC anticipates revisions to the NUREG as guidance in these other areas is developed.

In initiating this effort, NRC recognized that the Electric Power Research Institute (EPRI) also was performing work in this area with similar objectives. Both NRC and EPRI believed a collaborative effort to have technical agreement and to minimize duplication of effort would be more effective and efficient. NRC and EPRI have worked together under a Memorandum of Understanding and, as such, the two efforts complement each other.

TABLE OF CONTENTS – VOLUME 1
MAIN REPORT

TABLE OF CONTENTS -- VOLUME 1 (continued)

TABLE OF CONTENTS -- VOLUME 1 (continued)

LIST OF FIGURES – VOLUME 1

LIST OF TABLES – VOLUME 1

TABLE OF CONTENTS – VOLUME 2
APPENDICES[1]

LIST OF FIGURES – VOLUME 2

LIST OF TABLES – VOLUME 2

---

[1] Volume 2 of this report is being published at a later date. It is anticipated that the page numbering provided above may change; however, the page numbers for the Table of Contents, List of Figure and List of Tables that will be in Volume 2 will be accurate.

LIST OF TABLES (continued)

ACKNOWLEDGEMENTS

NRC and EPRI have collaborated in their mutual efforts on this topic. As such, both this NUREG and the EPRI documents complement each other. The NRC would like to acknowledge, that as part of this collaboration, EPRI (i.e., Ken Canavan) and its contractor ERIN Engineering (i.e., Don Vanover and Doug True) provided significant comments and developed the example provided in Appendix A of this NUREG.

ACRONYMS AND ABBREVIATIONS

ac	Alternating Current
ACRS	Advisory Committee on Reactor Safeguards
ALOCA	Large Break Loss of Coolant Accident
ANS	American Nuclear Society
AOT	Allowed Outage Time
ASME	American Society of Mechanical Engineers
ATWS	Anticipated Transient Without Scram
BWR	Boiling Water Reactor
CCF	Common Cause Failure
CDF	Core Damage Frequency
CFR	Code of Federal Regulations
CRD	Control Rod Drive
CST	Condensate Storage Tank
ECCS	Emergency Core Cooling System
EPRI	Electric Power Research Institute
ESW	Emergency Service Water
FCDF	Fire Core Damage Frequency
FW/PCS	Feedwater/Power Conversion System
HEP	Human Error Probability
HFE	Human Failure Event
HLR	High Level Requirement
HPCI	High Pressure Coolant Injection
HRA	Human Reliability Analysis
IA	Instrument Air
ICCDP	Incremental Conditional Core Damage Probability
ICLERP	Incremental Conditional Large Early Release Probability
IPEEE	Individual Plant Examination of External Events
ISLOCA	Interfacing System Loss of Coolant Accident
LAR	License Amendment Request
LERF	Large Early Release Frequency
LHS	Latin Hypercube Sampling
LLOCA	Large Break Loss of Coolant Accident
LOCA	Loss of Coolant Accident
LOOP	Loss of Offsite Power
LPSD	Low Power and Shutdown
LWR	Light Water Reactor
MCS	Minimal Cut Set
MLOCA	Medium Break Loss of Coolant Accident
MOV	Motor Operated Valve
MSIV	Main Steam Isolation Valve
NEI	Nuclear Energy Institute
NPP	Nuclear Power Plant
NRC	U.S. Nuclear Regulatory Commission
OOS	Out of Service
PCIG	Primary Containment Instrument Gas
PDF	Probability Density Function
POS	Plant Operating State

PRA	Probabilistic Risk Assessment
PWR	Pressurized Water Reactor
RAW	Risk Achievement Worth
RCIC	Reactor Core Isolation Cooling
RCP	Reactor Coolant Pump
RG	Regulatory Guide
RHR	Residual Heat Removal
RHRSW	Residual Heat Removal Service Water
RPV	Reactor Pressure Vessel
RVR	Reactor Vessel Rupture
RY	Reactor Year
SBO	Station Blackout
SER	Safety Evaluation Report
SGTR	Steam Generator Tube Rupture
SLOCA	Small Break Loss of Coolant Accident
SOKC	State of Knowledge Correlation
SPC	Suppression Pool Cooling
SR	Supporting Requirement
SSC	Structure System and Component
SSHAC	Senior Seismic Hazard Analysis Committee
TAF	Top of Active Fuel
YR	Year

1. INTRODUCTION

1.1 Background and History

In a 1995 policy statement [NRC, 1995], the U.S. Nuclear Regulatory Commission (NRC) encouraged the use of probabilistic risk assessment (PRA) in all regulatory matters. The policy statement declares that "the use of PRA technology should be increased to the extent supported by the state-of-the-art in PRA methods and data and in a manner that complements NRC's deterministic approach . . . PRA and associated analyses (e.g., sensitivity studies, uncertainty analyses and importance measures) should be used in regulatory matters . . . " The Commission further notes in the policy statement that the "treatment of uncertainty is an important issue for regulatory decisions. Uncertainties exist . . . from knowledge limitations . . . A probabilistic approach has exposed some of these limitations and provided a framework to assess their significance and assist in developing a strategy to accommodate them in the regulatory process."

In a white paper entitled, "Risk-Informed and Performance-Based Regulation" [NRC, 1999], the Commission defined the terms and described its expectations for risk-informed and performance-based regulation. The Commission indicated that a "risk-informed" approach explicitly identifies and quantifies sources of uncertainty in the analysis (although such analyses do not necessarily reflect all important sources of uncertainty) and leads to better decisionmaking by providing a means to test the sensitivity of the results to key assumptions.

Since the issuance of the PRA policy statement, NRC has implemented or undertaken many uses of PRA including modification of its reactor safety inspection program and initiation of work to modify reactor safety regulations. Consequently, confidence in the information derived from a PRA is an important issue. The technical adequacy of the content has to be sufficient to justify the specific results and insights to be used to support the decision under consideration. The treatment of the uncertainties associated with the PRA is an important factor in establishing this technical acceptability.

Regulatory Guide (RG) 1.200 [NRC, 2007a], RG 1.174 [NRC, 2002], and the national PRA consensus standard [ASME/ANS, 2009] each recognize the importance of the identification and understanding of uncertainties that are part of PRA (and of any deterministic analysis as well), and these references provide guidance on this subject to varying degrees. However, they do not provide explicit guidance on the treatment of uncertainties in risk-informed decisionmaking.

RG 1.200 states that a full understanding of the uncertainties and their impact is needed (i.e., sources of uncertainty are to be identified and analyzed). Specifically, RG 1.200 notes the following:

> "An important aspect in understanding the base PRA results is knowing what are the sources of uncertainty and assumptions and understanding their potential impact. Uncertainties can be either parameter or model uncertainties, and assumptions can be related either to PRA scope and level of detail or to model uncertainties. The impact of parameter uncertainties is gained through the actual quantification process. The assumptions related to PRA scope and level of detail are inherent in the structure of the PRA model. The requirements of the applications will determine whether they are acceptable. The impact of model uncertainties and related assumptions can be evaluated qualitatively or

quantitatively. The sources of model uncertainty and related assumptions are characterized in terms of how they affect the base PRA model (e.g., introduction of a new basic event, changes to basic event probabilities, change in success criterion, introduction of a new initiating event)."

RG 1.174 states that a PRA should include a full understanding of the impacts of the uncertainties through either formal quantitative analysis or more simple bounding or sensitivity analyses. The guidance also maintains that the decisions "must be based on a full understanding of the contributors to the PRA results and the impacts of the uncertainties, both those that are explicitly accounted for in the results and those that are not."

The national consensus standard on PRA requires that sources of model uncertainty in the base PRA be identified and provides requirements for the identification and characterization of both parameter and model uncertainties, both parameter and the model. However, the standard provides requirements on "what to do" and not "how to."

In a letter dated April 21, 2003 [ACRS, 2003a], the Advisory Committee on Reactor Safeguards (ACRS) provided recommendations for staff consideration in Draft Guide 1122 (now RG 1.200). One recommendation was to include guidance on how to perform sensitivity and uncertainty analyses. In its letter, the ACRS noted the following:

> ". . . a systematic treatment should include rigorous analyses for parametric uncertainties, sensitivity studies to identify the important epistemic uncertainties, and quantification of the latter. In a risk-informed environment, the proper role of sensitivity studies is to identify what is important to the results, not to replace uncertainty analyses."

In its response [NRC, 2003a], the staff noted that the standard provides requirements for the performance of sensitivity and uncertainty analyses. However, the staff agreed to examine the requirements in more detail to identify where additional guidance may be needed.

In a subsequent letter dated May 16, 2003 [ACRS, 2003b], the ACRS provided recommendations to improve the quality of risk information for regulatory decisionmaking. One recommendation was for the staff to develop guidance for the quantitative evaluation of model uncertainties. In its letter, the ACRS noted the following:

- If methods other than PRAs are used to compensate for missing scope items, they can result in nonconservative decisions.

- Models that are included in the PRAs can be important sources of uncertainty. For example, using only one of the several models for human performance yields results with unknown uncertainties because using another model could produce different results. Yet this model uncertainty is rarely considered.

- Most licensees have not included a systematic treatment of uncertainties in their PRAs. A systematic treatment would include analyses of parametric uncertainties, sensitivity studies to identify the important model uncertainties, and quantification of the latter.

- Tools for performing analyses of parametric uncertainties are readily available and are included in most of the widely-used PRA software. However, the disciplined use of sensitivity studies to address model uncertainties is not as well understood. Developing guidance for quantifying model uncertainty is feasible, and such an effort would build on past practice and the literature.

- More guidance regarding sensitivity and uncertainty analyses would contribute greatly to confidence in risk-informed regulatory decisionmaking. Such guidance should include a clear discussion of the roles of sensitivity and uncertainty analyses as well as practical procedures for performing these analyses. Guidance also should address how uncertainties should be treated in the PRA and how they impact decisionmaking. In addition, guidance should provide examples to show the pitfalls if uncertainties are inadequately addressed.

In response to the ACRS [NRC, 2003b], the Commission agreed that guidance is needed for the treatment of uncertainties in risk-informed decisionmaking (i.e., the role of sensitivities and uncertainty analyses), specifically guidance regarding acceptable characterization of other methods, such as bounding analyses, to ensure that reasonable approaches are used. This report provides the needed guidance.

1.2 Objectives

This document provides guidance on how to treat uncertainties associated with PRA in risk-informed decisionmaking. Specifically, guidance is provided with regard to:

- Identifying and characterizing the uncertainties associated with PRA.

- Performing uncertainty analyses to understand the impact of the uncertainties on the results of the PRA.

- Factoring the results of the uncertainty analyses into the decisionmaking.

With regard to the first two objectives, the American Society of Mechanical Engineers (ASME) and the American Nuclear Society (ANS) have been developing standards on PRA that support these objectives. Specifically, the ASME/ANS PRA standard [ASME/ANS, 2009] provides requirements[2] related to identifying, characterizing, and understanding the impact of the uncertainties. However, the standard only specifies what needs to be done. No guidance has been developed on how to meet these requirements or on how to factor the above into the decisionmaking. Consequently, the guidance provided in this document addresses these aspects.

The guidance provided in this report also is intended to be consistent with NRC's PRA policy statement and, subsequently, more detailed guidance in RG 1.200. In addition, the guidance is intended to support these documents and other NRC documents that address risk-informed activities including, at a minimum, the following:

- RG 1.174, "An Approach for Using Probabilistic Risk Assessment in Risk-Informed Decisions on Plant-Specific Changes to the Licensing Basis."

---

[2] The use of the word "requirement" is standards language (e.g., in a standard, it states that the standard "sets forth requirements") and the use of the word is not meant to imply a regulatory requirement.

- RG 1.201, "Guidelines for Categorizing Structures, Systems, and Components in Nuclear Power Plants According to Their Safety Significance" [NRC, 2006a].

- RG 1.205, "Risk-Informed, Performance-Based Fire Protection for Existing Light-Water Nuclear Power Plants" [NRC, 2006b].

- RG 1.206, "Combined License Applications for Nuclear Power Plants (LWR Edition)" [NRC, 2007b].

- Standard Review Plan, Section 19.0, "Probabilistic Risk Assessment and Severe Accident Evaluation for New Reactors," Revision 2; Section 19.1, "Determining the Technical Adequacy of Probabilistic Risk Assessment Results for Risk-Informed Activities;" Section 19.2, "Review of Risk Information Used to Support Permanent Plant-Specific Changes to the Licensing Basis: General Guidance" [NRC, 2007c].

- Regulatory guides for specific applications such as for inservice testing, inservice inspection, and technical specifications [NRC, 1998a, NRC, 2003c; NRC, 1998b].

The guidance also is intended to support guidance provided by standards-setting and nuclear industry organizations [NEI, 2005a; NEI, 2006]. In particular, the Electric Power Research Institute (EPRI), in parallel with NRC, has been developing guidance documents on the treatment of uncertainties. This work is meant to complement the guidance in this document. Where applicable, the NRC guidance will refer to the EPRI work for acceptable approaches for the treatment of uncertainties [EPRI, 2004; EPRI, 2006] (see Section 1.4).

1.3   Approach Overview

In implementing risk-informed decisionmaking, NRC expects that risk analyses and interpretation of results and insights will provide appropriate consideration of uncertainty. Such consideration should include using a program of monitoring, feedback, and corrective action to address significant uncertainties (i.e., those uncertainties that could potentially impact the decision). To meet this objective, it is necessary to understand the risk-informed decisionmaking process itself.

With the process defined, the characteristics of a risk model and, in particular, a PRA are described. This description includes recognition that a PRA, being a probabilistic model, characterizes aleatory uncertainty that results from randomness associated with the events of the model. The focus of this document is epistemic uncertainty (i.e., uncertainties in the formulation of the PRA model). Therefore, guidance is given on identifying and describing the different types of sources of epistemic uncertainty and the different ways that they are treated. The different types of epistemic uncertainty are parameter, model, and completeness uncertainties.

- Parameter Uncertainty. Guidance is provided on how to address parameter uncertainty in the use of PRA results for decisionmaking. This guidance involves characterization of parameter uncertainty, propagation of uncertainty, assessment of the significance of epistemic correlation, and comparison of results with acceptance criteria.

- **Model Uncertainty.** Guidance is provided for identifying and characterizing model uncertainties in PRAs. This guidance involves assessing the impact of model uncertainties on PRA results and insights used to support risk-informed decision.

- **Completeness Uncertainty.** Guidance is provided on addressing one aspect of completeness uncertainty (i.e., missing scope) in risk-informed applications. This guidance addresses the performance of a conservative or bounding analysis as one means to address items missing from a plant's PRA scope.

The ASME/ANS PRA Standard provides input to the decision process in that the standard (as endorsed by NRC) identifies what needs to be performed in understanding what sources of uncertainty are associated with a PRA. The guidance developed for each type of uncertainty relates also to the standard.

The final part of the guidance includes addressing the uncertainty in PRA results in the context of risk-informed decisionmaking. In particular, the guidance addresses interpretation of the results of the uncertainty analysis when comparing PRA results with the acceptance criteria established for a specified application.

1.4   Relationship to EPRI Report on Uncertainties

The NRC staff initiated work on the treatment of uncertainties as noted above. NRC recognized from the start that EPRI also was performing work in this area with similar objectives. Both NRC and EPRI believed a collaborative effort to have technical agreement and to minimize duplication of effort would be more effective and efficient. Consequently, NRC and EPRI agreed to work together under a Memorandum of Understanding to ensure the two efforts complemented each other. Figure 1-1 below shows the relationship between NRC work (i.e., this NUREG report) and EPRI work.

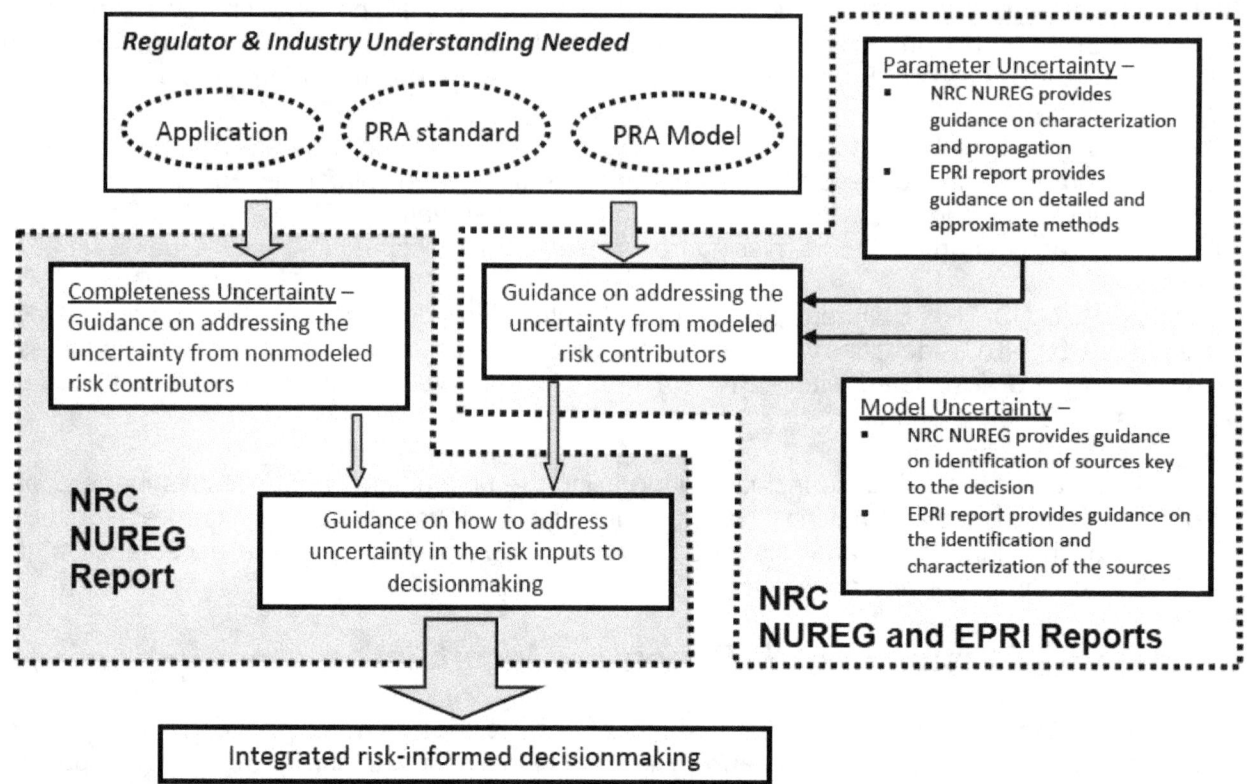

Figure 1-1 Relationship between NRC and EPRI efforts

In providing guidance on the treatment of uncertainties, both the NRC and EPRI work starts with (1) consideration of the decision under review, (2) the PRA standard, and (3) the supporting PRA model. However, NRC's focus is from a regulatory perspective while EPRI's focus is from an industry perspective. Both perspectives are essential when treating uncertainties in decisionmaking. These perspectives provide the context in development of the guidance for the different types of uncertainties. The uncertainties associated risk contributors modeled in the PRA include the parameter and modeled uncertainties. Both NRC and EPRI efforts provide guidance for these uncertainties. With regard to parameter uncertainties, this NUREG provides guidance on characterization and propagation while the EPRI report provides guidance on detailed and approximate methods. With regard to model uncertainties, this NUREG provides guidance on identification of sources key to the decision while the EPRI report provides guidance on the identification and characterization of the sources.

In addition, uncertainties exist from nonmodeled risk contributors (referred to as completeness uncertainty). This NUREG provides guidance on addressing these uncertainties.

So far, the guidance has focused on identifying and understanding the uncertainties associated with the risk contributors. This NUREG also provides the needed guidance on how to address the various uncertainties in the risk results and insights so that they are treated in the integrated risk-informed decisionmaking.

1.5 Scope and Limitations

The guidance in this document focuses on the use of PRA insights and results and ways to address the associated uncertainties. Consequently, the scope of the guidance contained in this report is limited to addressing the uncertainties associated with the use of the results of risk assessment models for risk-informed decisionmaking. Therefore, this document currently provides no guidance for the uncertainties associated with other analyses.

The current scope of the standard addresses internal and external events occurring during at-power conditions in estimating core damage frequency and large early release frequency. The standard currently does not address (1) low power and shutdown conditions, (2) the estimation of releases other than large early, and (3) the estimation of fatalities. Therefore, the guidance does not address the sources of uncertainty associated with those conditions. In addition, although the standard does include internal fire and external events, this guidance currently does not address the sources of uncertainty associated with internal fire and external events. This document addresses the sources of uncertainty for an at-power Level 1 and LERF PRA for internal events and internal flood.

Although the guidance does not currently address all sources of uncertainty, the guidance provided on the process for their identification and characterization and for how to factor the results into the decisionmaking is generic and is independent of the specific source. Consequently, the process is applicable for other sources such as internal fire, external events, and low power and shutdown.

In addressing uncertainties, expert judgment or elicitation may be used to determine if an uncertainty exists and the nature of the uncertainty (e.g., magnitude). This NUREG does not cover guidance for performing expert judgment or elicitation.

An expert panel may be convened for the decisionmaking to address those significant risk contributors not covered by a standard. Use of an expert panel will take into consideration the sources of uncertainty associated with those risk contributors. Guidance for employing an expert panel currently is not in the scope of the report.

In developing the guidance, the focus is on currently operating reactors. In particular, some of the numerical screening criteria referred to in this document and the identification of the sources of model uncertainty included in the EPRI document are informed by experience with PRAs for currently operating reactors. Although the process is applicable for advanced LWRs and non-LWRs and reactors in the design stage, the screening criteria and the specific sources of uncertainty may not be applicable. In addition, some sources unique to these reactors will exist that are not addressed in this report.

In developing the sources of model uncertainty, a model uncertainty needs to be distinguished from an assumption or approximation that is made, for example, on the needed level of detail. Although these assumptions and approximations can influence the decisionmaking, they are generally not considered to be model uncertainties because the level of detail in the PRA model could be enhanced, if necessary. Therefore, methods for addressing this aspect are not explicitly included in this report, and Section 5 discusses their consideration.

1.6   Report Organization

The remainder of this report is divided into the following sections:

- Section 2. Provides an overview of the overall approach used to address uncertainties in risk-informed decisionmaking, which includes the context of the uncertainties and an overview of the risk-informed decisionmaking process itself. This section serves as a roadmap to the rest of the report.

- Section 3. Defines the characteristics of a risk model (in particular, a PRA) and the sources and types of uncertainty.

- Section 4. Provides the guidance for the treatment of parametric uncertainty.

- Section 5. Provides the guidance for the treatment of model uncertainty.

- Section 6. Provides the guidance for the treatment of completeness uncertainty.

- Section 7. Provides guidance on addressing the uncertainty in PRA results in the context of risk-informed decisionmaking.

- Section 8. Provides the references.

- Appendix A. Provides a detailed example of how to apply the guidance in both this report and the EPRI document.

2. OVERALL APPROACH

This section provides an overview of the information and guidance provided in the subsequent sections of this report on addressing the treatment of uncertainties associated with probabilistic risk assessment (PRA) in risk-informed decisionmaking (RIDM). To understand and implement the guidance provided for the treatment of uncertainties, it is necessary to first comprehend what constitutes RIDM, the relationship of PRA to RIDM, the characteristics of PRA related to the associated sources of uncertainty, and the sources and their relationship to the PRA Standard. This understanding will establish the context needed to allow guidance to be developed on each type of uncertainty and to enable results or insights from these uncertainties to be implemented into RIDM.

2.1 Risk-Informed Decisionmaking Process

In its white paper, "Risk-Informed and Performance-Based Regulation" [NRC, 1999], the Commission defined a risk-informed approach to regulatory decisionmaking:

- A "risk-informed" approach to regulatory decisionmaking represents a philosophy whereby risk insights are considered together with other factors to establish requirements that better focus licensee and regulatory attention on design and operational issues commensurate with their importance to public health and safety.

This philosophy was elaborated in Regulatory Guide (RG) 1.174 [NRC, 2002] to develop a risk-informed decisionmaking process for licensing changes and has been implemented in U.S. Nuclear Regulatory Commission (NRC) risk-informed activities.

In developing this process, NRC defined a set of key principles in RG 1.174 to be followed for decisions regarding plant-specific changes to the licensing basis. For the most part, the following principles are global in nature and have been generalized to all activities that are the subject of risk-informed decisionmaking:

- Principle 1: Current Regulations Met.
- Principle 2: Consistency with Defense-in-Depth Philosophy.
- Principle 3: Maintenance of Safety Margins.
- Principle 4: Acceptable Risk Impact.
- Principle 5: Monitor Performance.

Of the five principles, Principle 4, Acceptable Risk Impact, is the one that most directly depends on the use of PRA. Consequently, Principle 4 is of most interest and, therefore, the guidance in this report focuses on an acceptable treatment of uncertainty in a risk assessment. However, to address the uncertainties associated with the risk assessment, a basic understanding regarding the use of all these principles in decisionmaking is needed.

The principles of RIDM are expected to be observed; however, they are not the process that is used in risk-informed decisionmaking. RG 1.174 presents an approach that ensures the principles will be met for RIDM involving plant-specific changes to the licensing basis. This approach can be generalized and applied to all risk-informed decisionmaking.

The generalized approach integrates all the insights and requirements that relate to the safety or regulatory issue of concern. These insights and requirements include recognition of any mandatory requirements resulting from current regulations as well as the insights from deterministic and probabilistic analyses performed to help make the decision. The generalized approach ensures that defense-in-depth measures and safety margins are maintained. It also includes provisions for implementing the decision and for monitoring the results of the decision. Figure 2-1 provides an illustration of this integrated process. The dotted line delineates those parts of the process that are addressed in this document.

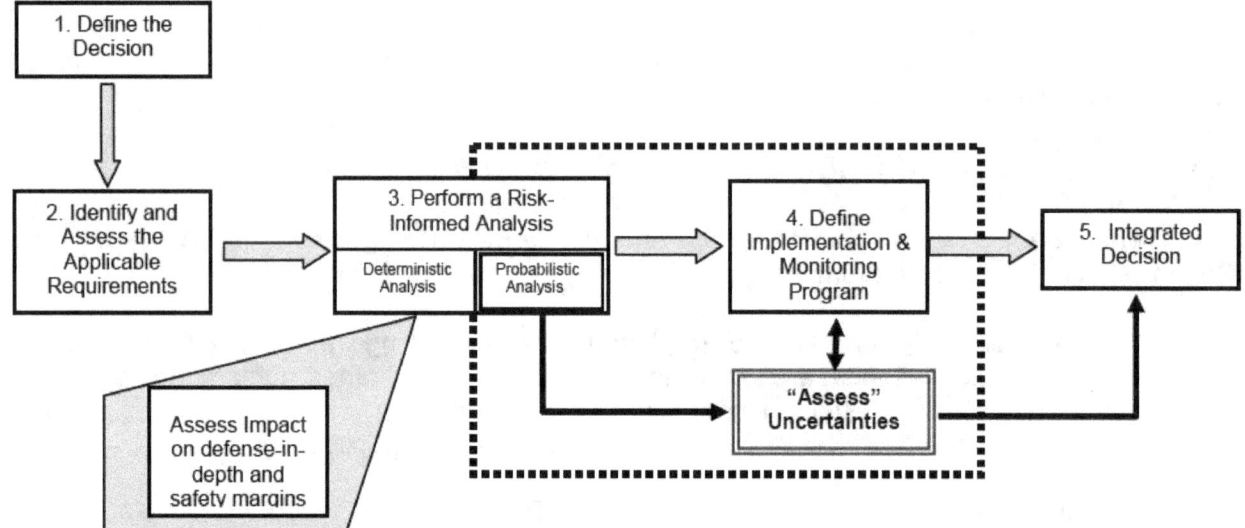

Figure 2-1 Elements of integrated risk-informed decisionmaking process

Element 1: Define the decision. The first step in the process is to define the issue or decision under consideration. Examples of the types of issues/decisions that NRC would need to address are related to:

• The design or operation of the plant.

• The plant technical specifications/limits and conditions for normal operation.

• The periodicity of inservice inspection, inservice testing, maintenance, and planned outages.

• The allowed combinations of safety system equipment that can be removed from service during power operation and shutdown modes.

• The adequacy of the emergency operating procedures and accident management methods.

Element 2: Identify and assess the applicable requirements. In this element, the current regulatory requirements that apply to the decision under consideration are identified. Part of this determination includes identifying (understanding) the effect of the applicable requirements on the decision. This element implements Principle 1 of risk-informed decision.

Element 3: Perform a risk-informed analysis. In this element, an assessment is made, in terms of a risk-informed analysis, to demonstrate that Principles 2, 3, and 4 are met. The risk-informed analysis includes both deterministic and probabilistic components. The deterministic component implements Principles 2 and 3, which take into account the impact on defense-in-depth and on safety margins. The probabilistic component implements Principle 4, acceptable risk impact. A treatment of the uncertainties in the probabilistic analysis is implicitly required to implement Principle 4 of risk-informed decisionmaking. Treatment of these uncertainties is the focus of this report.

Element 4: Define Implementation and Monitoring Program. In this element, a part of the decisionmaking process is understanding the effect of implementing a positive decision. This understanding involves determining how to monitor the change so that future assessments can be made as to whether the decision was implemented effectively and to guard against any unanticipated adverse effects. Consequently, consideration should be given to a performance-based means of monitoring the results of the decision.

Element 5: Integrated decision. In this final element of the decisionmaking process, the results from Elements 1 through 5 are integrated and the decision is made whether to accept or reject a proposed design, plant change, regulatory change, etc. This integration requires that the individual insights obtained from the other elements of the decisionmaking process are weighed and combined to reach a conclusion. An essential aspect of the integration is the consideration of uncertainties.

PRAs can address many uncertainties explicitly. These uncertainties are the epistemic uncertainties arising from recognized limitations in knowledge. However, a specific type of uncertainty exists that risk analyses, whether deterministic or probabilistic, cannot address. This type of uncertainty involves the incompleteness of the state of knowledge concerning potential failure modes or mechanisms. Because these failure modes or mechanisms are unknown, they cannot be addressed analytically (whether the analysis is deterministic or probabilistic). Principles 2, 3, and 5 (i.e., those related to defense-in-depth, safety margins, and performance monitoring) address the latter type of uncertainty. The focus of this report is on the treatment of the uncertainties associated with the PRA and the uncertainties regarding completeness with regard to the known risk contributors not modeled in the PRA.

To address the treatment of uncertainties associated with a PRA in the risk-informed decisionmaking process, it is necessary to understand the role the PRA plays in the decision, the different sources of uncertainties, and the impact of the uncertainties. Figure 2-2 shows this process, and the report provides the guidance for these various pieces.

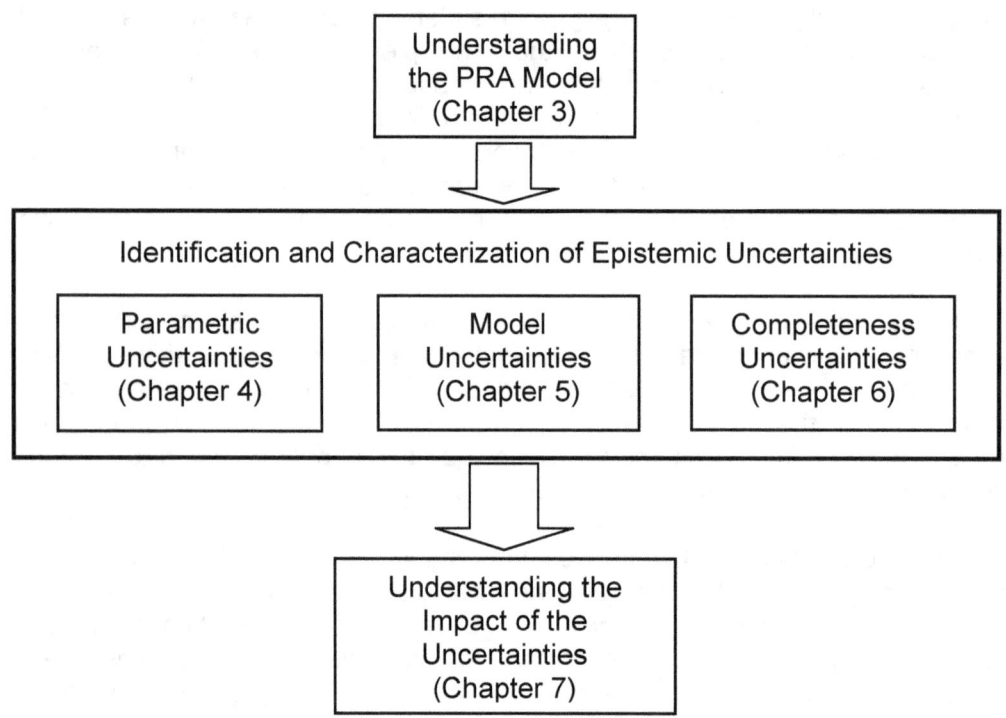

Figure 2-2 Steps needed for the treatment of uncertainties

The subsequent sections of this section provide an overview of the process for treating uncertainties; the later sections provide the detailed guidance.

2.2 Understanding the Risk Analysis Needed to Support a Decision

The purpose of Section 3 is to provide an overview of how PRA results are used in a risk-informed decision and to provide an overview of the characteristics of a PRA model. This overview is used to illustrate the origin and nature of the different sources of uncertainty associated with the PRA model that are the subject of this report. Part of this understanding includes what is explicitly specified in the PRA Standard regarding uncertainties.

A risk evaluation is required to support Element 3 of the decisionmaking process discussed in Section 2.1. The risk evaluation is performed by:

- Identifying the risk measures (e.g., core damage frequency [CDF]) to be used in evaluating the impact of the decision on risk.

- Identifying the numerical guidelines used to determine the acceptability of the risk impact.

- Constructing a PRA model to generate those results (i.e., the needed risk measures).

- Comparing the results to the acceptance guidelines.

- Documenting the conclusions.

When documenting the conclusions, it is necessary to address the robustness of the results taking into account the uncertainties in the risk assessment. One aspect of addressing the robustness of the results is determining whether the PRA model has sufficient detail to address the issues of concern and whether the level of detail has led to a bias on the results. However, the focus of this document is on the impact of uncertainties on the results.

To appropriately account for the uncertainties associated with the PRA results and insights in decisionmaking, a thorough understanding of the PRA model is required. This understanding will identify the underlying assumptions and limitations, thereby indicating the sources of uncertainty in the PRA results and insights. Understanding the sources of uncertainty will indicate the parts of the PRA model that could be affected and, ultimately, the results from the PRA model that may be impacted.

PRA models that address the risk from nuclear power plants are complex models, the development of which involves a number of different tasks. These tasks include the development of logic structures (e.g., event trees and fault trees) and the assessment of the frequencies and probabilities of the basic events of the logic structures. Both the development of the logic models and assessment of probabilities can introduce uncertainties that could have a significant impact on the predictions of the PRA model, and these uncertainties need to be addressed. Although uncertainties in a PRA model have different sources, the two basic classes of uncertainties are aleatory and epistemic [Apostolakis, 1994, Helton, 1996].

Aleatory uncertainty is associated with the random nature of events such as initiating events and component failures. The PRA model is an explicit model of the random processes and thus is a model of the aleatory uncertainty.

Epistemic uncertainties arise when making statistical inferences from data and, perhaps more significantly, from incompleteness in the collective state of knowledge about how to represent plant behavior in the PRA model. The epistemic uncertainties relate to the degree of belief that the analysts have in the representativeness or validity of the PRA model and in its predictions (i.e., how well the PRA model reflects the design and operation of the plant and, therefore, how well it predicts the response of the plant to postulated accidents).

The three types of epistemic uncertainties found in a PRA are parameter uncertainty, model uncertainty, and completeness uncertainty. The identification, understanding, and treatment of these three types of epistemic uncertainties found in a PRA are the principal subject of the remainder of this report.

2.3 Addressing Epistemic Uncertainty

The following is an overview of the treatment of the three different types of epistemic uncertainties (parameter, model, and completeness) as identified in Section 1.3. Sections 4, 5, and 6 provide detailed guidance on these three uncertainty types, respectively.

2.3.1 Parameter Uncertainty

Parameter uncertainty relates to the uncertainty in the computation of the input parameter values used to quantify the probabilities of the events in the PRA logic model. Examples of such parameters are initiating event frequencies, component failure rates and probabilities, and human error probabilities. These uncertainties can be characterized by probability distributions that relate to the analysts' degree of belief in the values of these parameters (which could be derived from simple statistical models or from more sophisticated models).

As part of the risk-informed decisionmaking process, the numerical results (e.g., CDF) of the PRA, including their associated parameter uncertainty, are compared with the appropriate decision criteria or guidelines. The uncertainties on the input parameters need to be combined in an appropriate manner to provide an assessment of this type of uncertainty on the PRA results. An important aspect of this propagation is the need to account for what has been called the state-of-knowledge correlation (SOKC) or epistemic correlation. This concern arises when the same parameter (including its uncertainty) is used to quantify the probabilities of two or more basic events. Most of the PRA software in current use has the capability to propagate parameter uncertainty through the analysis, taking into account the SOKC to calculate the probability distribution for the results of the PRA. In some cases, however, it may not be necessary to consider the SOKC. Section 4 examines the implications of the SOKC and provides guidance on when it is important to account for it.

2.3.2 Model Uncertainty

Model uncertainty arises because different approaches may exist to represent certain aspects of plant response and none is clearly more correct than another. Uncertainty with regard to the PRA results is then introduced because uncertainty exists with regard to which model appropriately represents that aspect of the plant being modeled. In addition, a model may not be available to represent a particular aspect of the plant. Uncertainty with regard to the PRA results is again introduced because there is uncertainty with regard to a potentially significant contributor not being considered in the PRA.

The uncertainty associated with the model and its constituent parts is typically dealt with by making assumptions. Examples of such assumptions include those made concerning (1) how a reactor coolant pump in a pressurized-water reactor would fail following loss-of-seal cooling, (2) the approach used to address common cause failure in the PRA model, and (3) the approach used to identify and quantify operator errors. In general, model uncertainties are addressed by determining the sensitivity of the PRA results to different assumptions or models.

The treatment of model uncertainty in decisionmaking depends on how the PRA will be used. There are two ways in which the PRA can be used:

(1)   The base PRA[3] is used as input to evaluating proposed changes to a plant's licensing basis (e.g., by using RG 1.174) and the emphasis is on the change in risk due to the proposed plant changes.

(2)   The base PRA is used in a regulatory application (e.g., to evaluate various design options or to determine the baseline risk profile as part of a license submittal for a new plant).

---

[3] The term "base PRA" is meant to denote the PRA that is developed to support the various applications; that is, the base PRA is independent of an application.

The expectation is that the focus in the decisionmaking will be on identifying and evaluating sources of model uncertainty that are key to the specific application at hand. Identifying the key sources of model uncertainties involves the following three steps:

(1)   Identification of Sources of Model Uncertainties and Related Assumptions of the Base PRA. Both generic and plant-specific sources of model uncertainty and related assumptions for the base PRA are identified and characterized. These sources of uncertainty and related assumptions are those that result from developing the PRA model.

(2)   Identification of Sources of Model Uncertainties and Related Assumptions Relevant to the Application. The sources of model uncertainty and related assumptions in the base PRA that are relevant to the application are identified. This identification may be performed with a qualitative analysis. This analysis is based on an understanding of how the PRA is used to support the application and the associated acceptance criteria or guidelines. In addition, new sources of model uncertainty and related assumptions that may be introduced by the application are identified.

(3)   Screening for Key Sources of Model Uncertainties and Related Assumptions for the Application. The sources of model uncertainty and related assumptions that are key to the application are identified. Quantitative analyses of the importance of the sources of model uncertainty and related assumptions identified in the previous steps are performed in the context of the acceptance guidelines for the application. The analyses are used to identify any reasonable alternative modeling hypotheses that could impact the decision. These hypotheses are used to identify which of the sources of model uncertainty and related assumptions are key.

Section 5 provides detailed guidance on the treatment of model uncertainty.

2.3.3 Completeness Uncertainty

Completeness uncertainty relates to risk contributors that are not in the PRA model. These types of uncertainties either are ones that are known but not included in the PRA model or ones that are not known and therefore not in the PRA model. Both types are important.

The known uncertainties (not included in the PRA model) could have a significant impact on the predictions of the PRA. Examples of sources of these types of incompleteness include the following:

*   The scope of the PRA does not include some classes of initiating events, hazards, or modes of operation.

*   The level of analysis may have omitted phenomena, failure mechanisms, or other factors because their relative contribution is believed to be negligible.

When a PRA is used to support an application, its scope and level of detail needs to be examined to determine if they match what is required for the risk-informed application. If the scope or level of detail of the existing base PRA is incomplete, then either the PRA is upgraded to include the missing piece(s) or it is demonstrated using conservative or bounding-type analyses that the missing elements are not significant risk contributors.

Section 6 focuses on this latter approach to the treatment of completeness uncertainty. The purpose of Section 6 is to provide the guidance for the performance of conservative or bounding analysis to address items missing from a plant's PRA scope or level of detail. However, this approach can only be used for those sources of incompleteness that have been recognized.

The second type of incompleteness uncertainties are the unknowns. Examples include the following:

- No agreement exists on how a PRA should address certain effects, such as the effects on risk resulting from aging or organizational factors. Furthermore, PRAs typically do not address them.

- The analysis may have omitted phenomena, failure mechanisms, or other factors because they are unknown.

These sources of incompleteness that are truly unknown are addressed in risk-informed decisionmaking by the other principles, such as safety margins, and discussed in Section 7.

2.4 Understanding and Assessing the Impact of the Uncertainties

In making an integrated decision, the decision maker needs to consider each of the elements in Figure 2-1. When the risk-informed analysis element is considered, the uncertainties associated with the risk analysis (i.e., the results) need to be addressed so that the robustness of the conclusions of the risk analysis is understood and appropriately considered in the decision.

The section stresses the need to understand the results of the risk assessment in detail. This understanding is required to identify the sources of uncertainty that are relevant to the decision. In addition, it is particularly important when the results are generated by combining results from PRA models for different hazard groups (e.g., internal initiating events, internal fires, seismic events). PRA models are developed to varying levels of detail for a number of reasons, and the influence of potentially conservative approximations made for ease of modeling, for example, needs to be identified as it can distort the results.

Depending on the significance of the uncertainties in the risk assessment results, the decision under consideration could be influenced. Section 7 provides guidance on addressing the uncertainty in PRA results in the context of risk-informed decisionmaking. In particular, guidance is provided on interpreting the results of the uncertainty analysis when comparing PRA results with the acceptance criteria or guidelines established for a specified application. In dealing with model uncertainty, the PRA analyst needs to provide the decisionmaker with an assessment of the credibility of any alternate models or assumptions that have the potential to alter the decision.

In other cases, an analyst may decide not to use a PRA model to address a contributor to risk. One way of quantitatively addressing the uncertainty in the PRA results that arises from using an incomplete PRA model is to use bounding approaches to demonstrate that a missing PRA scope contributor to risk is not significant to the decision. However, when the missing contributors cannot be demonstrated to be insignificant to the decision, an alternative approach is needed. One acceptable alternative is to impose conditions on the implementation of the decision to compensate for the uncertainty. These conditions would involve limiting the implementation such that the change does not affect the unquantified portion of the risk. Section 7 discusses this in more detail.

3. UNDERSTANDING THE RISK ASSESSMENT NEEDED TO SUPPORT A DECISION

This section provides (1) high-level guidance on defining the risk assessment needed to support a decision, (2) an overview of those features of the analysis that can have an impact on the confidence in the conclusions drawn from the analysis, (3) high-level discussion on the different types of sources of uncertainty that result from the modeling process and a brief description of how these uncertainties are addressed, and (4) Probabilistic Risk Assessment (PRA) Standard requirements regarding uncertainties defined in the American Society of Mechanical Engineers (ASME)/American Nuclear Society (ANS) PRA Standard [ASME/ANS, 2009].

3.1 Guidance on Defining the Required Risk Assessment

The principal steps for performing a risk assessment to support a decision are the following:

- Identify the results needed (acceptance guidelines).
- Construct a model to generate the required results.
- Compare the results to the acceptance guidelines.
- Document the results.

These steps are described in the following sections. In addition, the acceptance guidelines of the Nuclear Regulatory Commission's (NRC) Regulatory Guide (RG) 1.174 [NRC, 2002] are used to illustrate, in a concrete way, the ideas presented in this section.

3.1.1 Identify the Results Needed

When using the results of a risk assessment to support a risk-informed decision, the first step is to identify what results are needed and how they are to be used to inform the decision. The results needed are generally formulated in terms of acceptance guidelines or criteria (i.e., the results needed from the risk assessment should be organized in such a way that they can be compared to the acceptance guidelines associated with the risk-informed decision). For a regulatory application, the acceptance guidelines can be found in the corresponding regulatory guide or in industry documents that are endorsed by a regulatory guide. Acceptance guidelines can vary from decision to decision, but most likely they will be stated in terms of the numerical value or values of some risk metric or metrics. The metrics commonly used include:

- Core damage frequency (CDF).

- Large early release frequency (LERF).

- Change in core damage frequency (ΔCDF) or large early release frequency (ΔLERF).

- Conditional core damage probability (CCDP) or conditional large early release probability (CLERP).

- Incremental core damage probability (ICDP) or incremental large early release probability (ICLERP).

- Various importance measures such as Fussell-Vesely (FV), risk achievement worth (RAW), and Birnbaum.

Table 3-1 identifies the metrics associated with some common applications.

Table 3-1 Metrics used for specific applications.

Risk-informed applications	Acceptance Guidelines					Required Risk Metrics						
	Baseline CDF/LERF	Δ CDF/ΔLERF	CCDP/CLERP	ICCDP/ICLERP	FV/RAW importance	Annual average baseline CDF/LERF	Condition-specific annual average CDF/LERF	Condition-specific baseline CDF/LERF	Condition-specific CDF/LERF	Duration of condition	FV/RAW importance	Birnbaum importance
RG 1.174	x	x				x	x					
RG 1.175 – Inservice Testing	x	x			x	x	x				x	
RG 1.177 – Technical Specifications	x	x		x		x	x	x	x	x		
RG 1.178 – Inservice Inspection			x				x					
Technical Specification Initiative 5b	x	x				x	x					
Technical Specification Initiative 4b				x				x	x	x		
Maintenance Rule a(4)				x				x	x	x		
Notices Of Enforcement Discretion				x				x	x			
NRC Management Directive 8.3			x				x					
Significance Determination Process		x		x		x			x	x		
Title 10 Code of Federal Regulations Part 50.69					x	x	x				x	
Maintenance Rule Risk Significance					x						x	
Component Design Bases Inspection					x						x	
Component Risk Ranking (Motor Operated Valves, etc.)					x						x	
Mitigating System Performance Index		x				x	x				x	x

The acceptance guidelines also should include guidance on how the metric is to be calculated, in particular with regard to addressing uncertainty. Section 3.1.3 discusses this in more detail. In addition, when defining the metrics and the acceptance criteria, it is necessary to define the scope of risk contributors that should be addressed.

The scope of the risk assessment is defined in terms of the hazard groups, plant operational states, and risk metrics. Section 3.2 discusses this in more detail. For regulatory applications, the scope of risk contributors that needs to be addressed includes all hazard groups and all plant operational states that are relevant to the decision. For example, if the decision involves only the at-power operational state, then the low-power and shutdown operational states need not be addressed.

3.1.2 Construct a Model to Generate the Required Results

The next step is to determine how the required results are going to be generated. The generation of the results is typically done by using a quantitative risk model. Depending on the decision to be made, this assessment could be based on the baseline risk or it could be based on changes to the baseline risk. Guidance on how to perform the analysis may be provided in associated guidance documents for some of the well-defined applications, such as risk-informed in-service testing (IST). General guidance on using a PRA model in support of a risk-informed decision is provided in RG 1.174 (for changes to the licensing basis) and in the Probabilistic Safety Assessments (PSA) Applications Guide (Electric Power Research Institute [EPRI, 1995]). In these documents, the guidance focuses on establishing a cause-effect relationship to identify the portions of the PRA that are affected by the issue being evaluated and on determining how to quantify the impact on those PRA elements.

For example, a proposed application can impact one or more PRA technical elements. Examples of potential impacts include the following:

- Introducing a new initiating event or requiring modification of an initiating event group, such as all the loss-of-coolant accident (LOCA) initiators.

- Changing a system success criterion.

- Requiring the addition of new accident sequences.

- Requiring additional failure modes of structures, systems, and components (SSCs).

- Altering system reliability or changing system dependencies.

- Requiring modification of probabilities of basic events.

- Introducing a new common cause failure (CCF) mechanism.

- Eliminating, adding, or modifying a human action.

- Changing important results used in other applications, such as importance measures.

- Changing the potential for containment bypass or failure modes leading to a large early release.

- Changing the SSCs required to mitigate external hazards, such as seismic events.

- Changing the reliability of systems used during low-power and shutdown (LPSD) modes of operation.

In accordance with the Commission's Phased Approach to PRA Quality, the risk from each significant risk contributor (i.e., hazard group and/or plant operating state) should be addressed using a PRA model that is performed in accordance with a consensus standard for that risk contributor (hazard group or plant operating state) endorsed by the staff. A significant risk contributor is one whose consideration can make a difference to the decision. Contributors shown to be insignificant can be addressed by using conservative assessments or by screening.

3.1.3 Compare the Results to the Acceptance Guidelines

The comparison of risk assessment results to the acceptance guidelines is, in principle, straightforward. However, the robustness of the results need to be demonstrated if the decisionmaker is to have any confidence in the results of that comparison. To achieve the needed level of confidence, the results driving the conclusions need to be analyzed in detail to assess their realism and to identify and address the sources of uncertainty. The subsequent sections of this report focus on the approaches to addressing these issues. However, sections 3.2 and 3.3 provide an overview of those characteristics of PRA models that need to be addressed when drawing conclusions from the results of the risk assessment.

3.1.4 Document the Results

The results of the comparison with the acceptance guidelines are documented, and recommendations are provided to the decisionmakers. The analysts address the important sources of uncertainty and provide a statement characterizing their confidence in their recommendations.

3.1.5 An Example – RG 1.174

RG 1.174 is used to illustrate the issues associated with comparison of PRA results to acceptance criteria or guidelines in a risk-informed environment. The use of the RG 1.174 to illustrate the issues is driven by the fact that it is among the most widely used regulatory guide in the regulatory arena for nuclear reactors. Moreover, the formulation of the acceptance guidelines was informed by other acceptance criteria or guidelines in common use, such as those in the Regulatory Analysis Guidelines [NRC, 2004] and the subsidiary objective to the Safety Goal Policy Statement [NRC, 1986].

Identifying the Results Needed: The needed results are specified by the acceptance guidelines; for example, the guideline associated with the CDF metric is shown in Figure 3-1. In the context of RG 1.174, the intent of comparing the PRA results with the acceptance guidelines is to demonstrate with reasonable assurance that the change in risk associated with the decision is small (i.e., Principle 4 of the risk-informed decisionmaking process discussed in Section 2 is met). This decision needs to be based on a full understanding of the contributors to the PRA results and the impacts of the uncertainties, both those that are explicitly accounted for in the results and those that are not. This process is somewhat subjective, and the reasoning behind the decisions needs to be well documented.

Because of the way the guidelines were chosen, in particular the relationship to the Commission's subsidiary objectives for the safety goals, the metrics are such that, for a risk-informed decision, all contributors to risk should be addressed and each significant contributor to risk is taken into account. In this context, a significant contributor is one whose risk contribution can have an effect on the decision. The significant risk contributors define the scope of the risk assessment that need to be performed.

Constructing a Risk Model: RG 1.174 provides guidance on how to use the results of a PRA to support a decision. It recognizes that the PRA model may not be full scope and allows the use of alternate approaches to address the risk contributions from some hazard groups. However, the intent of the Commission's Phased Approach to PRA Quality is that the significant contributors be addressed using a PRA model. In other words, unless a risk contributor can be shown to be insignificant to the decision (i.e., having no impact on the decision), then it has to be addressed using a PRA model that meets an endorsed standard.

Since RG 1.174 is applicable to a general class of unspecified license amendments, the guidance provided on how to modify the PRA to generate the results is not explicit. Instead, the guidance addresses using a cause-effect relationship to identify the portions of the PRA being affected by the issue being evaluated and to determine how to quantify the impact on those PRA elements.

Section 2.2.5 of RG 1.174 discusses how the comparison of the calculated results with the guidelines is to be performed, particularly taking into account the uncertainties in the calculated results. This section includes a categorization of uncertainties into parameter uncertainty, model uncertainty, and completeness uncertainty.

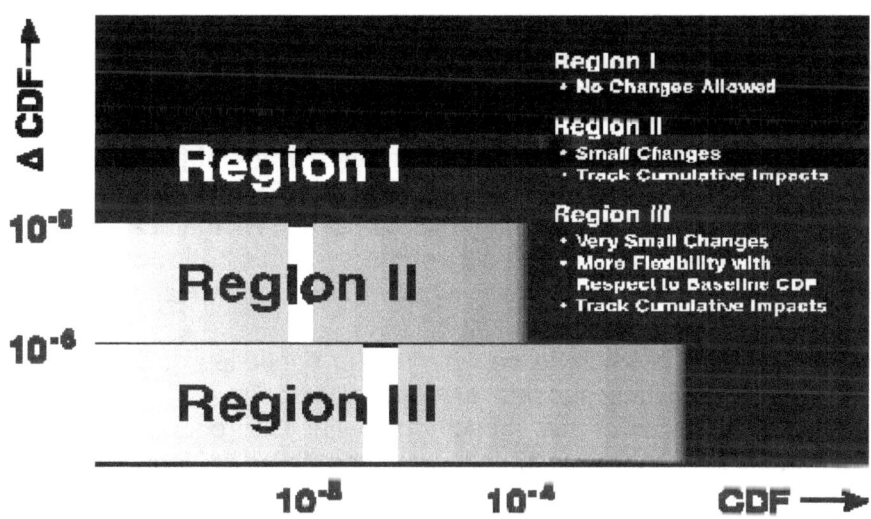

Figure 3-1 Example of risk acceptance guidelines

Comparison with the Guidelines: The acceptance guidelines of RG 1.174 are not intended to be interpreted as overly prescriptive. They are intended to provide an indication, in numerical terms, of what is considered acceptable. This guidance is largely in recognition that the state-of-knowledge, or epistemic, uncertainties associated with PRA calculations preclude a definitive conclusion, based purely on the numerical results, as to which of the three regions (shown in

Figure 3-1) the results belong. In particular, model uncertainties typically are not addressed in the uncertainty analysis used to calculate the mean values.

In addition, it is recognized that unquantified contributions may exist that could increase or decrease risk. In the case of RG 1.174, the term "acceptance guidelines" was used rather than "acceptance criteria" primarily to recognize that not all of the analysis uncertainty was captured in the uncertainty distribution. Section 7.6 discusses some implications of this distinction.

With respect to parameter uncertainty, the following two alternatives were discussed when establishing the acceptance guidelines for RG 1.174. The measure to be used for comparison should be (1) the mean value of the probability distribution representing the epistemic uncertainty or (2) a specified percentile of the uncertainty distribution, corresponding to a specified confidence level. The guidelines were finally chosen such that the appropriate measure for comparison is the mean value of the corresponding uncertainty distribution. The reason for this was to some extent historical because the guidelines were based on prior guidelines for which mean values were to be used. However, it also was considered that a philosophical problem exists associated with determining what would be the appropriate confidence level to choose if Option (2) above were adopted. More details of the reasoning behind the approach can be found in SECY-97-221 [NRC, 1997a].

3.2   Characteristics of a PRA Model

The characteristics of a PRA model are described below and include:

- Structure of the PRA model.
- Assumptions associated with the PRA model.
- Scope of the PRA model.
- Level of detail of the PRA model.
- PRA results from different hazard groups.

In general terms, a model can be described as an analyst's attempt to represent a "system." A system model in the physical sciences or engineering disciplines is usually a mathematical model, which is to say that it has a mathematical structure that can be used to produce numerical results that represent certain aspects of the system's behavior. Such a mathematical model will generally have several parameters, which require numerical estimates.

In general, because it is not possible to capture all the subtleties of the system behavior in a tractable mathematical form, most models are approximations. Therefore, uncertainties are associated with the formulation of the model and with its predictions. For some models, however, this uncertainty is so small that it can essentially be ignored. For example, the mathematical formulation of many of the models created by physicists to explain natural phenomena are well supported or verified such that the models are very precise in their predictions, and their uncertainty is sufficiently small that it can be ignored. An example of one such model is Newtonian mechanics and Newton's law of gravity. This model is capable of very accurately predicting such things as planetary motion and can be used to define the trajectories of planets or space vehicles with great accuracy.

PRA models are used to perform risk analysis of complex systems such as nuclear power plants. A PRA of a nuclear power plant is not as precise as the well-established physical models discussed in the above example for a number of reasons. For example, PRA models are probabilistic models that assess the probability or frequency of occurrence of rare events and their estimates cannot be verified directly. Furthermore, the phenomena arising from a severe accident and the equipment behavior under adverse conditions are not always well understood. These factors result in a PRA model involving varying degrees of approximation and based on a number of assumptions. The uncertainties associated with a PRA can range from being very small to having a significant effect on the calculated results.

To ensure that the decisionmaker is making an informed decision, an understanding of the uncertainties associated with the PRA model is essential. The analyst needs to understand the different kinds of uncertainties, identify the uncertainties, and then determine their impact on the results. Essential to this understanding is knowledge of the characteristics of a PRA model, including the structure of the model, the underlying assumptions made in developing the model, and the scope and level of detail of the model.

The analyst constructing the PRA model determines its scope and level of detail. Therefore, these two characteristics can vary to the extent that the PRA addresses significant contributors to a risk-informed decision. However, alternate approaches (other than inclusion in the PRA model) may be used for some risk contributors (whether they be specific hazard groups, specific initiating events, or specific accident sequences), particularly for items that are not significant contributors to the decision. In the context of risk-informed decisionmaking, these alternate approaches should be capable of producing estimates of risk and, therefore, are typically conservative or bounding models. Such alternate models (as discussed in Section 6) will have some characteristics of fully developed PRA models. Therefore, the discussion in Section 3.3 is applicable to the extent that the alternate model reflects the characteristics of a PRA model.

The characteristics of a PRA model include the structure of the model, the underlying assumptions made in developing the model, the scope of the model, the level of detail of the model, and the aggregation of the PRA model results. The following sections describe each of these characteristics in more detail.

3.2.1 The Structure of a PRA Model

A PRA model is a complex model consisting of many elements. The structural basis of the PRA model is a *logic model*, which is constructed using logic structures such as event trees and fault trees. The event trees identify the different plant responses in terms of sequences of undesired system states and human errors that could occur given an initiating event. The fault trees identify the different combinations of more elementary events (called basic events) that could lead to undesired system states. In combination, these logic structures capture the system dependencies such as the dependence of front-line systems on support systems. These logic models represent a simplification (discretization) of the potentially unlimited range of scenarios into a manageable set that is supposed to be representative of, and encompass the range of consequences of, that larger set in as realistic a way as practicable.

The types of basic events found in PRAs include events that represent the following:

- Occurrence of initiating events.

- The states of unavailability or failure of SSCs, and include failures due to common cause.

- The human failures that contribute to the unavailability or failure of the SSCs.

23

A frequency is estimated for each initiating event, while probabilities are estimated for the other basic events. These frequencies and probabilities are derived from the *basic event models*. For example, the occurrence of an initiating event is modeled as a random process with an associated frequency of occurrence. The other basic events discussed above are typically events such as the failure of a pump to start, the failure of a pump to run for 24 hours, common cause failure of a group of valves to change state, and the failure of an operator to take the appropriate actions to prevent system damage. Usually, these basic events also are regarded as resulting from random occurrences with respect to the demand created by the initiating event. They are described by simple probabilistic models, such as the constant probability of failure on demand (the binomial process) or a constant failure rate (the Poisson process).

Although the above examples are representative of most of the basic events in the models, other types of basic events also are included. For example, when analyzing reactor coolant pump (RCP) seal LOCAs, events may exist that delineate the different sizes of LOCAs. Furthermore, PRA models used to analyze the risk from fires or internal floods also may include events that are used to represent the occurrence of certain states of fire/flood damage. The probabilities of these events are derived from more complex and sometimes more uncertain models. For example, the probability of occurrence of a fire-damage state may be evaluated based on the complex interplay between a model of fire growth and a model of fire detection and suppression.

3.2.2 Assumptions in the PRA Model

The development of any model is generally based on a number of assumptions, and a PRA model is no exception. *An assumption* is a decision or judgment that is made in the development of the PRA model. An assumption is either related to a source of model uncertainty or related to scope or level of detail.

An *assumption related to a model uncertainty* is made with the knowledge that a different reasonable alternative assumption exists. A *reasonable alternative assumption* is one that has broad acceptance in the technical community and for which the technical basis for consideration is at least as sound as that of the assumption being made. Section 5 discusses this in more detail.

By contrast, an *assumption that is related to scope or level of detail* is one that is made for modeling convenience. Such assumptions result in defining the boundary conditions for the PRA model. Sections 3.2.4 and 3.2.5 address this topic in more detail.

3.2.3 Scope of a PRA Model

The scope of the PRA is defined in terms of (1) the metrics used to evaluate risk, (2) the plant operating states for which the risk is to be evaluated, and (3) the types of events (hazard groups) that can cause initiating events that challenge and disrupt the normal operation of the plant and, if not prevented or mitigated, would eventually result in core damage and/or a large release of radioactivity.

Risk metrics are the end-states (or measures of consequence) quantified in a PRA to evaluate risk. In a PRA, different risk metrics are generated by Level 1, limited Level 2, Level 2, or Level 3 PRA analyses.

- <u>Level 1 PRA</u>: Involves the evaluation and quantification of the frequency of the sequences leading to core damage. The metric evaluated is core damage frequency

- <u>Limited Level 2 PRA</u>: Involves the evaluation and quantification of the mechanisms and probabilities of subsequent radioactive material releases leading to large early releases from containment. The metric evaluated is the large early release frequency.

- <u>Level 2 PRA</u>: Involves the evaluation and quantification of the mechanisms, amounts, and probabilities of all the subsequent radioactive material releases from the containment. The metrics evaluated include the frequencies of different classes of releases, which include large release, early release, large late release, small release.

- <u>Level 3 PRA</u>: Involves the evaluation and quantification of the resulting consequences to both the public and the environment from the radioactive material releases. The metrics are typically measures of public risk that include frequencies of early fatalities and latent cancer fatalities.

Plant operating states (POSs) are used to subdivide the plant operating cycle into unique states such that the plant response can be assumed to be the same for all subsequent accident-initiating events. Operational characteristics (such as reactor power level; in-vessel temperature, pressure, and coolant level; equipment operability; and changes in decay heat load or plant conditions that lead to different success criteria) are examined to identify those relevant to defining POSs. These characteristics are used to define the states, and the fraction of time spent in each state is estimated using plant-specific information. The risk perspective is based on the total risk associated with the operation of the nuclear power plant, which includes not only full-power operation but also other operating states such as low-power and shutdown conditions.

Initiating events perturb the steady state operation of the plant by challenging plant control and safety systems whose failure could potentially lead to core damage and/or radioactivity release. These events include failure of equipment from either internal plant causes (such as hardware faults, operator actions, floods, or fires) or external plant causes (such as earthquakes or high winds). The challenges to the plant are classified into ***hazard groups***, which are defined as a group of similar causes of initiating events that are assessed in a PRA using a common approach, methods, and likelihood data for characterizing the effect on the plant. Typical hazard groups for a nuclear power plant PRA include internal events, seismic events, internal fires, internal floods, and high winds.

The scope of the model may be limited on the determination that a certain hazard group does not significantly affect the risk, as discussed in Section 6. On the other hand, the analyst may simply choose not to model a certain contributor and to deal with it in another manner, as discussed in Section 7.

3.2.4 Level of Detail

The level of detail of a PRA is defined in terms of the degree to which (1) the potential spectrum of scenarios is discretized and (2) the actual plant is modeled. Ultimately, the degree of detail required of the PRA is determined by how it is intended to be used. Although the goal of a PRA is to be as realistic as practicable, some compromise with realism is necessary as discussed below.

The logic models of a PRA (i.e., the event trees and fault trees) are a simplified representation of the complete range of potential accident sequences. For example, modeling all the possible initiating events or all the ways a component could fail would create an unmanageable and unwieldy model. Consequently, simplifications are achieved by making approximations. As an example, initiating events are consolidated into groups whose characteristics bound the characteristics of the individual members. As another example, when developing an accident sequence timeline, a representative sequence is generally chosen that assumes that all the failures of the mitigating systems occur at specific times (typically the time at which the system is demanded). However, in reality, the failures could occur over an extended time period (e.g., the system could fail at the time demanded or could fail at some later time). Developing a model that represents all the possible times the system could fail and the associated scenarios is not practical. The time line is used, among other purposes, to provide input to the human reliability analysis. Typically, a time is chosen that provides the minimum time for the operator to receive the cues and to complete the required action. This minimized time maximizes the probability of failure. This simplification, therefore, leads to an uncertainty in the evaluation of risk that is essentially unquantifiable without developing more detailed models that more explicitly model different timelines. The assumption is that the simplification is adequate for the purpose for which the model is being developed. It also is generally assumed that the simplification results in a somewhat conservative assessment of risk.

The degree to which plant performance is represented in the PRA model also has an effect on the precision of the evaluation of risk. For each technical element of a PRA, the level of detail may vary by the extent to which the following occur:

- Plant systems and operator actions are credited in modeling the plant design and operation.

- Plant-specific experience and the operating history of the plant's SSCs are incorporated into the model.

- Realism is incorporated in the deterministic analyses to predict the expected plant responses.

The level of detail in the way the logic models are discretized and the extent to which plant representation is modeled is at the discretion of the analyst. The analyst may screen out initiating events, component failure modes, and human failure events so that the model does not become encumbered with insignificant detail. For example, not all potential success paths may be modeled. However, a certain level of detail is implicit in the requirements of the ASME/ANS PRA Standard. Although an analyst chooses the level of detail, the PRA model needs to be developed enough to correctly model the major dependencies (e.g., those between front line and support systems) and to capture the significant contributors to risk. Nonetheless, the coarser the level of detail, the less precise is the estimate, resulting in uncertainty about the predictions of the model. The generally conservative bias that results could be removed by developing a more detailed model.

In many cases, the level of detail will be driven by the requirements of the application for which the PRA is being used. In particular, the PRA model needs to adequately reflect the cause-effect relationship associated with an application. As an example, typically PRAs are modeled on the assumption of annual average initiating event frequencies. Some applications, however, may require a more detailed breakdown. For example, a Notice of Enforcement Discretion (NOED) being requested during a time of severe grid stress might not be well reflected by an annual average estimate; indeed, the annual average might underestimate the risk. On the other hand, the assumption of annual average service water requirements may be conservative for an application that will be implemented only in the winter time.

3.2.5 Combining PRA Results from Different Hazard Groups (Aggregation)

Typically, the results produced by the PRA model include the following:

- Risk metric(s), such as CDF and/or LERF.

- Identification of the relative importance of various contributors including
    — Hazards considered.
    — Plant operational states.
    — Initiating events.
    — Accident sequences.
    — Component failures or unavailabilities.
    — Human failure events.

For many applications, it is necessary to consider the contributions from several hazard groups and/or plant operational states to a specific risk metric such as CDF, LERF, or an importance measure. Because the hazard groups and plant operating states are independent, addition of the contributions is mathematically correct. However, several issues should be considered when combining PRA results. First, it is important to note that, when combining the results of PRA models for several hazard groups (e.g., internal events, internal fires, seismic events) as required by many acceptance criteria, the level of detail and level of approximation may differ from one hazard group to the next with some being more conservative than others. This modeling difference is true even for an internal events, at-power PRA. For example, the evaluation of room cooling and equipment failure thresholds can be conservatively evaluated leading to a conservative time estimate for providing means for alternate room cooling. Moreover, at-power PRAs follow the same general process as used in the analysis of other hazard groups with regard to screening: low risk sequences can be modeled to a level of detail sufficient to prove they are not important to the results.

Significantly higher levels of conservative bias can exist in PRAs for external hazards, LPSD, and internal fire PRAs. These biases result from several factors, including the unique methods or processes and the inputs used in these PRAs as well as the scope of the modeling. For example, the fire modeling performed in a fire PRA can use simple scoping models or more sophisticated computer models or a mixture of methods and may not mechanistically account for all factors such as the application of suppression agents. Moreover, in an effort to reduce the number of cables that have to be located, fire PRAs do not always credit all mitigating systems. To a certain level, conservative bias will be reduced by the development of detailed models and corresponding guidance for the analysis of external hazards, fires, and LPSD that will provide a similar level of rigor to the one currently used in internal events at-power PRAs. However, as with internal events at-power PRAs, the evaluation of some aspects of these other contributors will likely include some level of conservatism that may influence a risk-informed decision.

The level of detail, scope, and resulting conservative biases in a PRA introduces uncertainties in the PRA results. Because conservative bias can be larger for external events, fire, and LPSD risk contributors, the associated uncertainties can be larger. However, a higher level of uncertainty does not preclude the aggregation of results from different risk contributors; but it does require that sources of conservatism having a significant impact on the risk-informed application be recognized.

Finally, it is important to note that the process of aggregation can be influenced by the type of risk-informed application. For example, it is always possible to add the CDF (LERF), or the changes in CDF (LERF) contributions from different hazard groups for comparison against corresponding criteria. However, in doing so, one should always consider the influence of known conservatism when comparing the results against the criteria, particularly if they mask the real risk contributors (i.e., distort the risk profile) or result in exceeding the criteria. If the criteria are exceeded due to a conservative analysis, then it may be possible to perform a more detailed, realistic analysis to reduce the conservatism and uncertainty. For applications that use risk importance measures to categorize or rank SSCs according to their risk significance (e.g., revision of special treatment), a conservative treatment of one or more of the hazard groups can bias the final risk ranking. Moreover, the importance measures derived independently from the analyses for different hazard groups can not be simply added together and thus would require a true integration of the different risk models to evaluate them.

Section 7 provides guidance on how to address the aggregation of the results from different risk contributors for use in risk-informed decisionmaking. To facilitate this effort, it is best that results and insights from all of the different risk contributors relevant to the application be provided to the decisionmaker in addition to the aggregated results. This information will allow for consideration of at least the main conservatisms associated with any of the risk contributors and will help focus the decisionmaker on those aspects of the analysis that have the potential to influence the outcome of the decision.

3.3   PRA Models and Uncertainty

PRA models are constructed as probabilistic models to reflect the random nature of the constituent basic events such as initiating events and component failures. Randomness is one manifestation of a form of uncertainty that has come to be called *aleatory uncertainty*. Therefore, a PRA is a probabilistic model that characterizes the aleatory uncertainty associated with accidents at nuclear power plants (NPPs). The focus of this document is guidance in dealing with another area of uncertainty, namely *epistemic uncertainty*. This uncertainty is associated with the incompleteness in the analysts' state of knowledge about accidents at NPPs and has an impact on the results of the PRA model. The following sections identify and describe examples of sources of epistemic uncertainty, types of epistemic uncertainty, and assessments of the impact of uncertainty.

3.3.1 Examples of Epistemic Uncertainty in PRA

Uncertainties in the PRA models arise for many different reasons, including the following:

- Generally accepted probability models exist for many of the basic events of the PRA model. These models are typically simple, with only one or two parameters. Examples include the simple constant failure rate reliability model, which assumes that the failures of a component while it is on standby occur at a constant rate, and the uniformly distributed (in time) likelihood of an initiating event. The model for both these processes is the well-known Poisson model. The parameter(s) of such models may be estimated using appropriate data, which in the examples above comprise the number of failures observed in a population of like components in a given time and the number of occurrences of a fire scenario in a given time, respectively. Statistical uncertainties are associated with the estimates of the parameters of the model. Because most of the events that constitute the building blocks of the risk model (e.g., some initiating events, operator errors, and equipment failures) are relatively rare, the data are scarce and the uncertainties can be relatively significant.

- For some events, while the basic probability model is generally agreed on, uncertainties may be associated with interpreting the data to be used for estimation. For example, when collecting data on component failures from maintenance records, it is not always clear whether the failure would have prevented the component from performing the mission required of it to meet the success criteria assumed in the risk model.

- For some basic events, uncertainty can exist as to how to model the failures, which results in uncertainties in the probabilities of those failures. One example is the behavior of RCP seals on loss of cooling. Another example is the modeling of human performance and the estimation of the probabilities of human failure events.

- Uncertainty can exist about the capability of some systems to perform their function under the conditions specified by the developed scenarios. This leads to uncertainty in characterizing the success criteria for those functions, which has an impact on the logic structure of the model. One example is uncertainty about the capability of the components in one room of the NPP to perform their functions after loss of cooling to the room.

As seen in these examples, the uncertainty associated with the structure of and input to the PRA model can be in the choice of the logic structure, in the mathematical form of the models for the basic events, or in the values of the parameters of those models, or it can be in both. To the extent that changes in parameter values are little more than subtle changes in the form of the model, it can be argued that no precise distinction exists between model uncertainty and parameter uncertainty. However, as discussed below, parameter uncertainties and model uncertainties are dealt with differently. In addition, it should be noted that, while the Poisson and binomial models are typically adopted for the occurrence of initiating events and for equipment failures, using these models may not be appropriate for all situations.

3.3.2 Types of Epistemic Uncertainty

As noted above, it is helpful to categorize uncertainties into those that are associated with the parameter values and those that involve aspects of models. This categorization is primarily because the methods for the characterization and analysis of uncertainty are different for the two types. However, a third type of uncertainty exists, namely uncertainty about the completeness of the model. Although this type of uncertainty cannot be handled analytically, it needs to be considered when making decisions using the results of a risk assessment.

Parameter Uncertainty: Parameter uncertainty is the uncertainty in the values of the parameters of a model given that the mathematical form of that model has been agreed to be appropriate. Current practice is to characterize parameter uncertainty using probability distributions on the parameter values [SNL, 2003]. When the parameters are combined algebraically to evaluate the PRA numerical results or some intermediate result such as a basic event probability, these uncertainty distributions can be mathematically combined in a simple way to estimate the uncertainty of those numerical results. When uncertainty exists as to which model to use to estimate the probability of the basic events, this is more appropriately addressed as a model uncertainty.

Model Uncertainty: Model uncertainty is related to an issue for which no consensus approach or model exists and where the choice of approach or model is known to have an effect on the PRA model (e.g., introduction of a new basic event, changes to basic event probabilities, change in success criterion, and introduction of a new initiating event). A model uncertainty results from a lack of knowledge of how SSCs behave under the conditions arising during the development of an accident. A model uncertainty can arise for the following reasons:

- The phenomenon being modeled is itself not completely understood (e.g., behavior of gravity-driven passive systems in new reactors, or crack growth resulting from previously unknown mechanisms).

- For some phenomena, some data or other information may exist, but it needs to be interpreted to infer behavior under conditions different from those in which the data were collected (e.g., RCP seal LOCA information).

- The nature of the failure modes is not completely understood or is unknown (e.g., digital instrumentation and controls).

As indicated in Section 3.3.1, model uncertainty may be manifested in uncertainty about the logic structure of the PRA model or in the choice of model to estimate the probabilities associated with the basic events.

Completeness Uncertainty: Lack of completeness is not in itself an uncertainty, but recognition of the limitations in the scope of the model. However, the result is an uncertainty about where the true risk lies. This uncertainty also can be thought of as a type of model uncertainty. However, completeness uncertainty is discussed separately because it represents a type of uncertainty that cannot be quantified and because it represents those aspects of the system that are, either knowingly or unknowingly, not addressed in the model.

The problem with completeness uncertainty in PRAs is that, because it reflects an unanalyzed contribution, it is difficult (if not impossible) to estimate its magnitude. Incompleteness in the model can be thought of as arising in two different ways:

- Some contributors/effects may be knowingly left out of the model for a number of reasons. For example, methods of analysis have not been developed for some issues, and these gaps have to be accepted as potential limitations of the technology. Thus, for example, the impact on actual plant risk from unanalyzed issues such as the influences of organizational performance cannot now be explicitly assessed. As another example, the resources to develop a complete model may be limited, which could lead to a decision not to model certain contributors to risk (e.g., seismically induced fires).

- Some phenomena or failure mechanisms may be omitted because their potential existence has not been recognized.

3.3.3 Assessing the Impact of Uncertainty

Although many sources of uncertainty may exist in a PRA, they do not all have an impact on a particular decision. Those sources that can affect the results used to support a decision are identified as key sources of uncertainty, and these demand attention. The way that the uncertainties are assessed is a function of the way the acceptance criteria (or acceptance guidelines) for the decision are defined and the way that the uncertainties are characterized with regard to the acceptance guidelines.

For example, in the context of RG 1.174 [NRC, 2002], the acceptance guidelines are defined to demonstrate, with reasonable assurance, that the risk change is small. This decision has to be based on a full understanding of the contributors to the PRA results and the impacts of both the uncertainties that are explicitly accounted for in the results and those uncertainties that are not. This process is somewhat subjective, and the reasoning behind the decisions needs to be well documented. The acceptance guidelines of RG 1.174, for example, are not intended to be interpreted as being overly prescriptive. They are intended to indicate in numerical terms what is considered acceptable. RG 1.174 uses the term "acceptance guidelines" rather than "acceptance criteria" primarily to recognize that the numerical results do not capture all of the analysis uncertainty.

The following provides an overview of the assessment of each type of uncertainty (parameter, model, and completeness). Detailed guidance appears in Sections 4, 5, and 6, respectively.

Parameter Uncertainty: For many parameters (e.g., initiating event frequencies, component failure probabilities or failure rates, human error probabilities), the uncertainty may be characterized as subjective probability distributions. Section 4 discusses the methods for propagating these uncertainties through the PRA model to characterize the uncertainty in the numerical results of the analysis. In this manner, the impact of the parameter uncertainties on the numerical results of the PRA can be assessed integrally. However, many of the acceptance criteria or guidelines used in regulatory decisionmaking (e.g., the acceptance guidelines of RG 1.174) are defined such that the appropriate measure for comparison is the mean value of the uncertainty distribution on the corresponding metric. In this case, as discussed in Section 4, the primary issue with parameter uncertainty is its effect on the calculation of the mean, and specifically, on the relevance and significance of the state-of-knowledge correlation.

Model Uncertainty: Although the analysis of parameter uncertainty is fairly mature and is addressed adequately through the use of probability distributions on the values of the parameters, the analysis of the model and completeness uncertainties cannot be handled in such a formal manner. The typical response to a modeling uncertainty is to choose a specific modeling approach to be used in developing the PRA model. Although it is possible to embed a characterization of model uncertainty into the PRA model by including several alternate models and providing weights (probabilities) to represent the degree of credibility of the individual models, this approach is not usual. Notable exceptions are NUREG 1150 [NRC 1990] and the approach to seismic hazard evaluation proposed by the Senior Seismic Hazard Analysis Committee (SSHAC) [LLNL, 1997]. The approach taken in this document is, when using the results of the PRA model, it is necessary to demonstrate that the key uncertainties, reasonable alternative hypotheses, or modeling methods would not significantly change the assessment. Section 5 discusses methods for performing such demonstrations.

In dealing with model uncertainties, it is helpful to identify whether the model uncertainties can alter the logic structure of the PRA model or whether their primary impact is on the probabilities of the basic events of the logic model. For example, an uncertainty associated with the establishment of the success criterion for a specific system can result in an uncertainty as to whether one or two pumps are required for a particular scenario. This uncertainty would be reflected in the choice of the top gate in the fault tree for that system. On the other hand, those model uncertainties associated with choosing the model to estimate the probabilities of the basic events do not alter the structure of the model. Because of this, tools such as importance analyses can be used to explore the potential impact of these uncertainties in a way not possible for those uncertainties related to the logic structure of the PRA model.

One approach to dealing with a specific model uncertainty is to adopt a consensus model that essentially removes the uncertainty related to the choice of model from having to be addressed in the decisionmaking. In the context of regulatory decisionmaking, a consensus model can be defined as follows:

> *Consensus model.* In the most general sense, a model that has a publicly available published basis and has been peer reviewed and widely adopted by an appropriate stakeholder group. In addition, widely accepted PRA practices may be regarded as consensus models. Examples of the latter include the use of the constant probability of failure on demand model for standby components and the Poisson model for initiating events. For risk-informed regulatory decisions, the consensus model approach is one that NRC has utilized or accepted for the specific risk-informed application for which it is proposed.

The definition given here ties the consensus model to a specific application. This restriction is because models have limitations that may be acceptable for some uses and not for others. In some cases (e.g., the Westinghouse Owners' Group 2000 RCP seal LOCA model), this consensus is documented in a safety evaluation report (SER) [NRC, 2003d]. In many cases, the tendency is for the model to be considered somewhat conservative. If this is the case, it is important to recognize the potential for this conservatism to mask other contributors that may be important to a decision. Many models can already be considered consensus models without the issuance of an SER. For example, the Poisson model for initiating events has been used since the very early days of PRA.

It should be noted that adoption of a consensus model does not indicate that uncertainty is not associated with its use. However, this uncertainty would be manifested as an uncertainty on the results used to generate the probability of the basic event(s) to which the model is applied. An example of this approach is discussed in section 4 of a paper by Zio and Apostolakis [Zio, 1996]. This would be treated in the PRA quantification as a parameter uncertainty. What is removed by adopting a consensus model is the need to consider other models as alternatives.

Completeness Uncertainty: The issue of completeness of scope and level of detail of a PRA can be addressed for those scope and level of detail items for which methods are available and, therefore, some understanding of the contribution to risk exists. For example, the out-of-scope and level-of-detail items can be addressed by supplementing the PRA with additional analysis to enlarge the scope or increase the level of detail or by bounding analyses to demonstrate that, for the application of concern, the out-of-scope or beyond-the-level-of-detail contributors are not significant. Section 6 discusses these approaches. Section 7 discusses an alternative to design the proposed change such that the major sources of uncertainty will not have an impact on the decisionmaking process.

The true unknowns (i.e., those related to issues whose existence is not recognized) cannot be addressed analytically. However, in the interests of making defendable decisions, these unknowns are addressed during the decisionmaking as discussed in Section 2. The principles of safety margins and defense in depth play a critical role in addressing this source of uncertainty.

3.3.4 Summary

The following sections of the report describe approaches for addressing the parameter, model, and incompleteness uncertainties, respectively, and how to address these uncertainties in the results of a PRA that is being used to support a risk-informed decision.

In this context, the following should be noted:

- A PRA model is developed using a number of simplifying assumptions and approximations that are made for modeling convenience.

- It is expected that a PRA model being used to support a risk-informed decision been developed to the level of detail required to support that decision.

- The methods discussed here are not intended to address any biases that result from the modeling approximations and simplifying assumptions.

- Instead, these modeling approximations and simplifying assumptions are taken as defining the boundary conditions for the model.

- The uncertainties being addressed are those associated with:
  — The parameters used to quantify the probabilities of the basic event of the PRA logic model.
  — The choice of modeling approach.
  The completeness of scope of the model.

3.4 ASME/ANS PRA Standard Supporting Requirements

In understanding (1) the risk analysis needed to support a decision, (2) those features of the analysis that can have an impact on the confidence in the conclusions drawn from the analysis, and (3) the different types of sources of uncertainty that result from the modeling process, it also is important to know the PRA Standard addresses uncertainty.

Uncertainty is addressed either directly or indirectly in many of the supporting requirements (SRs) in the ASME/ANS PRA Standard and is related to either the quantification of a basic event or of a risk metric. SRs that are directly relevant include the following:

Basic Events

- IE-C1 and IE-C13 for the quantification of initiating event frequencies.

- HR-D6 for the quantification of pre-initiating event human error probabilities (HEPs).

- HR-G9 for the quantification of post-initiating event HEPs.

- DA-D1 and DA-D3 for the estimation of the parameters of basic events related to hardware failures or unavailability.

Risk Metric

- QU-A3 for the estimation of CDF.
- QU-E1 thru E4 for the identification and characterization of uncertainty in CDF.
- QU-F4 for the documentation.
- LE-E4 for the estimation of LERF.
- LE-F3 for the characterization of uncertainty in LERF.

Table 3-2 gives the requirement of each of the above SRs as stated in the mentioned addendum. Consistent with the Standard, boldface is used in this table to highlight the differences among the requirements in the three Capability Categories[4].

Table 3-2 SRs of the ASME/ANS PRA Standard related to uncertainty.

ASME/ANS Standard RA-Sa-2009	Description of the capability category...		
	I	II	III
IE-C1	CALCULATE the initiating event frequency accounting for relevant generic and plant specific data unless it is justified that there are adequate plant specific data to characterize the parameter value and its uncertainty. (See also IE-C13 for requirements for rare and extremely rare events.)		
IE-C15	CHARACTERIZE the uncertainty in the initiating event frequencies and PROVIDE mean values for use in the quantification of the PRA results.		
HR-D6	PROVIDE an assessment of the uncertainty in the HEPs in a manner consistent with the quantification approach. USE mean values when providing point estimates of HEPs.		

---

[4] The Standard is intended for a wide range of applications that require a corresponding range of PRA capabilities. Although the range of capabilities required for each portion of the PRA to support an application falls on a continuum, the Standard defined three Capability Categories so that requirements can be developed and presented in a manageable way. They are designated as PRA Capability Categories (CCs) I, II, and III. In general, the SRs are differentiated by the CCs. In other words, the specific stipulations of a SR may be different depending on the CC.

Table 3-2 SRs of the ASME/ANS PRA Standard related to uncertainty.

ASME/ANS Standard RA-Sa-2009	Description of the capability category...		
	I	II	III
HR-G8	Characterize the uncertainty in the estimates of the HEPs in a manner consistent with the quantification approach, and PROVIDE mean values for use in the quantification of the PRA results.		
DA-D1	USE plant-specific parameter estimates for events modeling the unique design or operational features if available, or use generic information modified as discussed in DA-D2; USE generic information for the remaining events.	CALCULATE realistic parameter estimates for significant basic events based on relevant generic and plant specific evidence unless it is justified that there are adequate plant specific data to characterize the parameter value and its uncertainty. When it is necessary to combine evidence from generic and plant specific data USE a Bayes update process or equivalent statistical process that assigns appropriate weight to the statistical significance of the generic and plant specific evidence and provides an appropriate characterization of uncertainty, CHOOSE prior distributions as either noninformative or representative of variability in industry data. CALCULATE parameter estimates for the remaining events by using generic industry data.	CALCULATE realistic parameter estimates based on relevant generic and plant specific evidence unless it is justified that there are adequate plaʃ""nt specific data to characterize the parameter value and its uncertainty. When it is necessary to combine evidence from generic and plant specific data USE a Bayes update process or equivalent statistical process that assigns appropriate weight to the statistical significance of the generic and plant specific evidence and provides an appropriate characterization of uncertainty. CHOOSE prior distributions as either non-informative, or representative of variability in industry data.

Table 3-2 SRs of the ASME/ANS PRA Standard related to uncertainty.

ASME/ANS Standard RA-Sa-2009	Description of the capability category...		
	I	II	III
DA-D3	PROVIDE a characterization (e.g., qualitative discussion) of the uncertainty intervals for the estimates of those parameters used for estimating the probabilities of the significant basic events.	PROVIDE a mean value of, and a statistical representation of the uncertainty intervals for, the parameter estimates of significant basic events. Acceptable systematic methods include Bayesian updating, frequentist method, or expert judgment.	PROVIDE a mean value of, and a statistical representation of the uncertainty intervals for, the parameter estimates. Acceptable systematic methods include Bayesian updating, frequentist method, or expert judgment.
QU-A3	ESTIMATE the point estimate CDF.	ESTIMATE the mean CDF accounting for the state-of-knowledge correlation between event probabilities when significant [Note (1)]. [5]	CALCULATE the mean CDF by propagating the uncertainty distributions, ensuring that the state-of-knowledge correlation between event probabilities is taken into account.
QU-E1	IDENTIFY sources of model uncertainty		
QU-E2	IDENTIFY assumptions made in the development of the PRA model.		
QU-E3	ESTIMATE the uncertainty interval of the CDF results. Provide a basis for the estimate consistent with the characterization of parameter uncertainties. (DA-D3, HR-D6, HR-G8, IE-C15).	ESTIMATE the uncertainty interval of the CDF results. ESTIMATE the uncertainty intervals associated with parameter uncertainties (DA-D3, HR-D6, HR-G8, IE-C15), taking into account the state-of-knowledge correlation.	PROPAGATE parameter uncertainties (DA-D3, HR-D6, HR-G8, IE-C15), and those model uncertainties explicitly characterized by a probability distribution using the Monte Carlo approach or other comparable means. PROPAGATE uncertainties in such a way that the state-of-knowledge correlation between event probabilities is taken into account.
QU-E4	For each source of model uncertainty and related assumptions identified in QU-E1 and QU-E2, respectively, IDENTIFY how the PRA model is affected (e.g., introduction of a new basic event, changes to basic event probabilities, change in success criterion, introduction of a new initiating event) [Note (1)][6].		
LE-E4	QUANTIFY LERF consistent with the applicable requirements of Tables 2-2.7-2(a), 2-2.7-2(b), and 2-2.7-2(c).[7]		

---

[5] NOTE 1 to this SR in the standard provides a brief description of this correlation.

[6] Note 1 to this SR in the standard states that "For specific applications, key assumptions and parameters should be examined both individually and in logical combinations."

[7] There is a general note to the LE-E SRs in the standard which states "*The supporting requirements in these tables are written in CDF language. Under this requirement, the applicable quantification requirements in Table 2-2.7-2 should be interpreted based on the approach taken for the LERF model. For example, supporting requirement QU-A2 addresses the calculation of point*

Table 3-2 SRs of the ASME/ANS PRA Standard related to uncertainty.

ASME/ANS Standard RA-Sa-2009	Description of the capability category...		
	I	II	III
LE-F3	IDENTIFY and CHARACTERIZE the LERF sources of model uncertainty and related assumptions, in a manner consistent with the applicable requirements of Tables 2-2.7-2(d) and 2-2.7-2(e).[8]		

Guidance on how to achieve each of the above SRs is provided in the subsequent sections, where appropriate.

---

estimate/mean CDF. Under this requirement, the application of QU-A2 would apply to the quantification of point estimate/mean LERF.

[8] There is a general note to the LE-F SRs in the standard which states "The supporting requirements in these tables are written in CDF language. Under this requirement, the applicable quantification requirements in Table 2-2.7 should be interpreted based on LERF, including characterizing the sources of model uncertainty and related assumptions associated with the applicable contributors from Table 2-2.8-3. For example, supporting requirement QU-D6 addresses the significant contributors to CDF. Under this requirement, the contributors would be identified based on their contribution to LERF.

4. PARAMETER UNCERTAINTY

This section provides guidance on addressing parameter uncertainty in probabilistic risk assessment (PRA) results. The guidance focuses on performing parameter uncertainty analyses in the context of the requirements defined in the American Society of Mechanical Engineers (ASME)/American Nuclear Society (ANS) PRA Standard [ASME/ANS, 2009]. Although the standard does not address parameter uncertainties for importance measures, this section provides additional guidance for the parameter uncertainty associated with these measures. It is vital to present the results of the uncertainty analysis in a manner that can be factored into the decisionmaking process. Consequently, this section provides guidance with regard to:

- Meeting the Supporting Requirements (SRs) related to parameter uncertainty of the ASME/ANS PRA Standard.

- Assessing parameter uncertainty of importance measures.

- Providing input on parameter uncertainty to the decisionmaker.

4.1 ASME/ANS PRA Standard Supporting Requirements Related to Parameter Uncertainty

Parameter uncertainty is addressed either directly or indirectly in many of the SRs in the ASME/ANS PRA Standard and is related to either the quantification of a basic event or of a risk metric. SRs that are directly relevant include the following:

Basic Events

- IE-C1 and IE-C15 for the quantification of initiating event frequencies.

- HR-D6 for the quantification of pre-initiating event human error probabilities (HEPs).

- HR-G8 for the quantification of post-initiating event HEPs.

- DA-D1 and DA-D3 for the estimation of the parameters of basic events related to hardware failures or unavailability.

Risk Metric

- QU-A3 for the estimation of core damage frequency (CDF).
- QU-E3 for the identification and characterization of uncertainty in CDF.
- LE-E4 for the estimation of large early release frequency (LERF).

Table 3-2 in Section 3 provides the detailed definition of each of the SRs.

The guidance below is separated between the basic events and the risk metrics.

39

4.1.1 Parameter Uncertainty of Basic Events

This section addresses the SRs for IE-C1, IE-C15, HR-D6, HR-G8, DA-D1, and DA-D3.

These SRs in the Standard use words such as "calculate frequency," "characterize the uncertainty," "provide (or use) mean values," and "provide a mean value of, and a statistical representation of the uncertainty intervals." Although different words are used, the intent of the various SRs is the same. The Standard requires (1) the probability of a basic event to be calculated and (2) the uncertainty associated with the parameters of the basic event to be characterized. It also should be noted that some SRs in the Standard use the term "uncertainty interval," which this report interprets to mean a characterization of the uncertainty. This characterization could, in the simplest approach, take the form of an interval (i.e., a range of values within which the value lies). However, it is more usual to characterize the uncertainty in terms of a probability distribution on the value of the quantity of concern, whether it is a parameter accident sequence frequency or a core damage frequency.

The probability of a basic event (or frequency of an initiating event) is calculated using a probability model, which will be referred to here as a "basic event model" that may have one or more parameters. A simple example of a basic event model is the exponential distribution for the failure times of a component, which has a single parameter, λ, the failure rate.

The term "parameter uncertainty of a basic event" is used in this report to mean the uncertainty in the probability or frequency of a basic event due to the uncertainty in the parameter(s) associated with the corresponding basic event model. The choice of the basic event model itself is subject to model uncertainty, which is discussed in Section 5. The uncertainty in the parameter is the subject of this section.

The parameter values are estimated based on available relevant data, which are relatively scarce for many of the parameters. Therefore, the uncertainty associated with the parameter estimates can be relatively large, which is one of the reasons it needs to be addressed.

The types of basic events referred to in the ASME/ANS PRA Standard for which parameter uncertainties need to be characterized include the following:

- IE-C1 and IE-C15:
  — Initiating events.

- HR-D6 and HR-G8:
  — Human failure events.

- DA-D1 and DA-D3:
  — Failures to start or change state of components.
  — Failures to run or maintain state of components.
  — Unavailabilities of components from being out-of-service.
  — Common-cause failures (CCFs) of components.
  — Failures to recover structures, systems, and components (SSCs), such as failure to recover offsite power within a certain time.
  — Other relevant failures, sometimes called special events.[9]

---

[9] An example of other relevant failures is the failure of the containment in a LERF PRA.

The PRA Standard is intended for a wide range of applications that require a corresponding range of PRA capabilities (i.e., the Standard recognizes that, for some applications, the level of detail, the level of plant specificity, and the level of realism needed in the PRA are commensurate with the intended application). Consequently, the Standard defines three PRA Capability Categories (CCs) that are meant to support the range of applications.

The three CCs are distinguished by the extent to which (1) the plant design, maintenance, and operation are modeled, (2) plant-specific information with regard to SSC performance/history is incorporated, and (3) realism with regard to plant response is addressed. Generally, from CC I to CC III, the level of detail, plant specificity, and realism increase.

For IE-C1, IE-C15, HR-D6, and HR-G8, no distinction exists among the capability categories; however, for DA-D1 and DA-D3, the requirements differ for each capability category as described below.

- IE-C1, IE-C15, HR-D6, and HR-G8. These SRs have the same requirements for the three CCs. In other words, they specify that the uncertainty in the parameters of the basic events (initiating events, human failure events, and recovery events) need to be characterized regardless of the application. These SRs are not attempting to specify the levels of characterization but only state that characterization of the uncertainty is needed. Consequently, different levels of rigor were not established in the Standard for these SRs. However, different levels of rigor can be applied in characterizing the uncertainty, and the different levels needed become apparent in the Standard in subsequent and related SRs. Specifically, the SRs associated with calculation of the risk metrics (i.e., QU-A3, QU-E3, LE-E4, and LE-F3 provide the different levels of rigor that need to be kept in mind for IE-C1, IE-C15, HR-D6, and HR-G8).

- DA-D1 and DA-D3 These SRs have different requirements for the three CCs. Unlike IE-C1, IE-C15, HR-D6, and HR-G8, the different levels of rigor for characterizing the uncertainty in the parameters for the basic events associated with component failures are specified in the SR rather than in subsequent and related SRs. Consequently, DA-D1 and DA-D3 specify that (1) characterization of the uncertainty is needed and (2) the level of rigor is needed for the characterization.

In characterizing the uncertainty of a basic event, an analyst is characterizing the uncertainty of each parameter used to calculate the probability of the basic event (i.e., the parameters of the basic event model). The approach for characterizing the uncertainty of each parameter depends on the level of rigor needed for the characterization (i.e., on the three CCs), as follows:

- Capability Category I. To satisfy CC I, point estimates (usually mean values) of the parameters of the basic events are provided. In addition, the uncertainty of the parameters may be characterized, for example, by specifying a range of values or, as mentioned by SR DA-D3, by giving a "qualitative discussion" of the range of uncertainty.

- Capability Category II. The Standard uses the term "uncertainty interval" for characterizing parameter uncertainty. To satisfy CC II, however, the intent of the Standard is inferred to be that the parameter estimates need to be represented as probability distributions. This characterization will enable the evaluation of a mean value and will provide a statistical representation of the uncertainty in the value of the parameter. Moreover, this characterization will allow the propagation of this uncertainty characterization through the PRA model so that the mean value of the PRA results (e.g., CDF, accident sequence frequency) and a quantitative characterization of their uncertainty can be generated.

- Capability Category III. To meet CC III, the parameter estimates have to be represented as probability distributions, as in the case of CC II. The distinction in the Standard between CC II and CC III is related to the significance of the basic event that each applies to and not to characterizing uncertainty. For example, for DA-D3, CC II requires only significant basic events to be included, while CC III requires all the basic events to be included.

SR DA-D3 identifies three acceptable methods for characterizing the parameter uncertainty of the parameters of the basic events (for CC II and CC III): (1) the frequentist method, (2) Bayesian updating, and (3) expert judgment. Details can be found in a number of statistical texts and in the *Handbook of Parameter Estimation for Probabilistic Risk Assessment* [SNL, 2003] for the frequentist and Bayesian methods. For expert judgment, details can be found in NUREG/CR-6372 [LLNL, 1997] or NUREG-1563 [NRC, 1996]. These methods are briefly discussed below.

- Frequentist Approach. The frequentist, or classical, approach defines the probability of a random event as the long-term fraction of times that the event would occur in a large number of trials. The frequentist approach typically provides a point estimate, the most common of which is the maximum likelihood estimate, and provides confidence bounds at specified levels of confidence.

- Bayesian Approach. The Bayesian approach characterizes what is known about the parameter in terms of a probability distribution that measures the current state of belief in the possible values of the parameter. The mean value of this distribution is typically used as the point estimate for the parameter. The Bayesian approach provides a formal approach for combining different data sources.

  The SRs for QU and LE for CC II and CC III essentially require that the Bayesian approach be used (instead of the frequentist method) for at least the parameters of the significant basic events. Table 4-1 summarizes the reasons (slightly modified from Appendix B of the *Handbook of Parameter Estimation for Probabilistic Risk Assessment*, which identifies the characteristics of these methods for assessing the uncertainty of the results of a PRA). Subsection 4.1.2, Step 5, discusses the propagation of the input parameter distributions to the evaluation of CDF and LERF.

Table 4-1   Characteristics of two methods for assessing parameter uncertainty.

Item	Bayesian approach	Frequentist approach
Characterization of probability distribution	Provides a probability distribution on the parameter value.	Provides a confidence interval that cannot be interpreted directly as a probability that the parameter lies in the interval.
Propagation of parameter uncertainty through PRA model	Bayesian distributions can be propagated through fault trees, event trees, and other logic models.	It is difficult or impossible to propagate frequentist confidence intervals through fault and event tree models common in PRA to generate the corresponding estimates of intervals on the output quantities of interest.

For the reasons mentioned above, the frequentist approach is not used commonly for assessing the uncertainty of a parameter as part of a PRA. However, it is often used when a rough estimate of a parameter is all that is required.

- Expert Judgment. The expert judgment approach relies on the knowledge of experts in the specific technical field who arrive at "best estimates" of the distribution of the probability of a parameter or basic event. This approach is typically used when detailed analyses or evidence concerning the event represented by a basic event are very limited or unavailable. Such a situation is usual in studying rare events. Ideally, this approach provides a mathematical probability distribution with values of a central tendency of the distribution (viz., the mean) and of the dispersion of the distribution, such as the 5th and 95th percentiles. The distribution represents the expert or "best available" knowledge about the probability of the parameter or basic event. The process of obtaining these estimates typically is called "expert judgment elicitation," or simply "expert judgment" or "expert elicitation."

An important example of expert judgment elicitation is the estimates obtained for nine issues, such as the reactor coolant pump (RCP) seal loss-of-coolant accident (LOCA) during the development of NUREG-1150 [NRC, 1990]. NUREG/CR-4550, Vol. 2 [SNL, 1989] discusses the results of expert judgment on these issues.

Conceptually, the guidance provided in this section applies to all the types of basic events listed in Section 4.1 above. The following references provide more specific guidance for CCFs and human failure events:

CCFs of Components

- NUREG/CR-5497 [INL, 1998a].
- NUREG/CR-6268 [INL, 1998b].
- NUREG/CR-5485 [INL, 1998c].
- NUREG/CR-4780 [EPRI, 1988].
- EPRI NP-3967 [EPRI, 1985].

<u>Human Failure Events</u>

- NUREG-1792 [NRC, 2005].

The parameter uncertainty in the estimates of the probabilities of failures to recover SSCs and of special events, such as the probability of RCP seal failure, should be addressed explicitly in the documents addressing each of them. In some cases, however, the study of these kinds of events provides an estimate of the probability of an event without giving the associated probability distribution. For example, the uncertainties in the probabilities of recovering alternating current power after a loss of offsite power are often not evaluated. In these cases, an estimate of the probability distributions should be included so the parameter uncertainty associated with a particular event is included in the PRA model.

In addition, the probabilities of some failures to recover SSCs and of special events are derived from quite complex models. For instance, in the fire PRA, the probability of a fire scenario is obtained from the combination of models of fire growth and suppression phenomena. As described above, each probability should have an associated parameter uncertainty. On the other hand, uncertainty also exists about the predictions of specific models used for the evaluations. However, this uncertainty would be manifested as an uncertainty on the results used to generate the probability of the basic event(s) to which the model is applied. Section 4 of a paper by Zio and Apostolakis [Zio, 1996] discusses an example of this approach that would be treated in the PRA quantification as a parameter uncertainty. Uncertainty related to the choice of model, however, would be treated as a model uncertainty as discussed in Section 5 of this document.

4.1.2 Evaluation of Risk Metric and Associated Probability Distribution

This section addresses the SRs for QU-A3, QU-E3, and LE-E4. These SRs, for assessing the CDF or the LERF, differ depending on the CC. Hence, meeting the SRs related to estimating a risk metric and associated probability distribution can be achieved using different approaches, as described below.

When addressing the estimation of the risk metric of CDF or LERF (SR QU-A3), CC I requires only the point estimate. For CCs II and III, the SR requires that the mean for a risk metric is provided. For CCs II and III, the mean is calculated ensuring that the epistemic correlation (EC) (the ASME/ANS PRA Standard refers to it as the state-of-knowledge correlation) is taken into account. However, for CC II, the evaluation of the mean may ignore this correlation if it is demonstrated that this correlation is not significant for the particular case being evaluated.

This correlation arises because, for identical or similar components in a given nuclear power plant, the state of knowledge about their failure parameters is the same. Appendix A of this section discusses the correlation in more detail.

When addressing the uncertainty interval or probability distribution (SR QU-E3), CC I requires only an estimate of the uncertainty interval and its basis. For CCs II and III, the probability distribution (mentioned in the Standard as uncertainty interval) for a risk metric is obtained by propagating the parameter uncertainty using the Monte Carlo method, or similar means, through the PRA model. The probability distribution is evaluated ensuring that the epistemic correlation is taken into account. However, for CC II, this correlation may be ignored in calculating the probability distribution provided that it is shown not to be significant for the particular case under assessment.

44

Accordingly, the approach for meeting the SRs for quantifying a risk metric and the associated parameter uncertainty for CC I is less rigorous than those required for CCs II and III:

- For CC I, the approach consists of two steps: (1) evaluating the PRA model to generate the point estimate of a risk metric, and (2) estimating the uncertainty interval of the risk metric.

- For CC II, the approach takes one of two paths, depending on whether the EC is significant. The first one is to generate the cutsets of the PRA model by evaluating the point estimate of a risk metric; the cutsets are used in the later step of propagating parameter uncertainty. The second is entering the parameter uncertainty data for each basic event into the PRA model. The third is defining the groups of basic events that are correlated due to the EC. The fourth is determining whether the EC is significant for the particular PRA model being evaluated. In this step, if the EC is found to be not significant for the particular case being evaluated, the step of setting up the groups of correlated basic events in the PRA code is skipped and the process moves on to the final step of propagating parameter uncertainty. In the fourth step, if the EC is found to be significant, the groups of correlated basic events defined in the third step are set up in the PRA code. The final step is to carry out an uncertainty evaluation to propagate parameter uncertainty through the PRA model.

- For CC III, the approach is essentially the same as the one for CC II. However, for the former, the step of setting up the groups of correlated basic events in the PRA code is always carried out, ensuring that the EC is accounted for.

Figure 4.1 presents an overview of the approach for each CC. In general, the steps of the approaches for CCs II and III do not necessarily have to be performed in the order proposed here. For example, Steps 1 and 2 may be carried out in any order. The order of the steps presented here is offered only as a general guideline. Moreover, it is assumed that a PRA model was implemented in a PRA computer code.

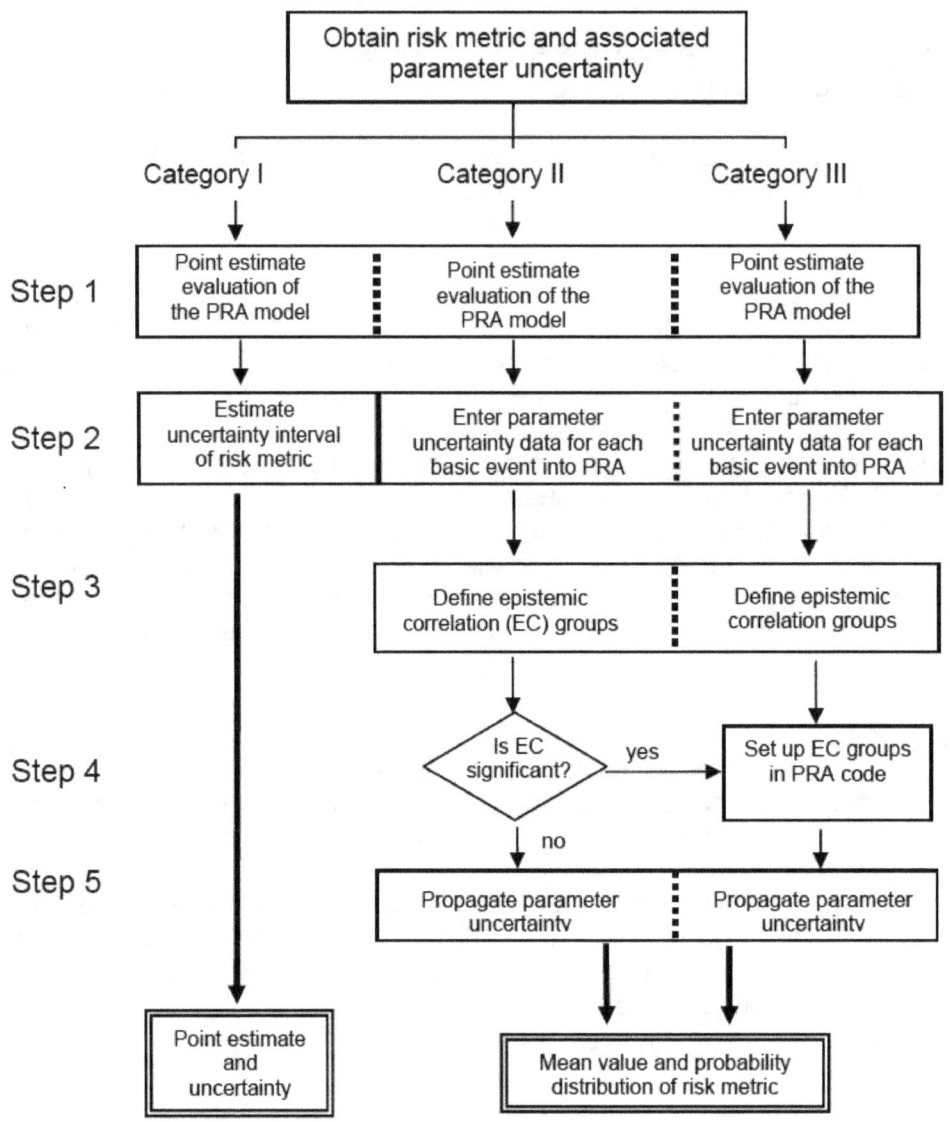

Figure 4-1   Approach for obtaining the mean value and parameter uncertainty of a risk metric

46

Step 1: Evaluate Point-estimate of the PRA Model

A solution of the PRA model yields the cutsets of the logic model of the PRA. The PRA computer code then can generate a point estimate of the CDF or LERF by quantifying these cutsets. It employs the point estimates of the basic events to obtain the point estimate of the CDF or LERF. In accordance with SRs IE-C15, HR-D6 and HR-G8, and DA-D3, for CCs II and III, the point estimate of each basic event, including each initiating event, should equal the mean value.

This activity always is completed for all CCs. For CCs II and III, the cutsets obtained from this activity procedure are used subsequently to propagate parameter uncertainty throughout the logic model of the PRA, as described in Step 5.

Step 2 (CC I): Estimate Uncertainty Interval of Risk Metric

This activity is undertaken only for CC I. In this case, the SR QU-E3 indicates that the uncertainty interval of the risk metric is estimated and that a basis for the estimate, consistent with the characterization of parameter uncertainties, should be provided. This estimation and its basis are case-specific and, for this reason, no generic guidance is offered.

At this point, the SRs for QU-A3, QU-E3, and LE-E4 for CC I have been met. The remaining step pertains to CCs II and III for these SRs.

Step 2 (CCs II and III): Enter Parameter Uncertainty Data for Basic Events into PRA Code

This activity is undertaken only for CCs II and III. For each basic event in the PRA model, the information about the probability distribution of each of its parameters is entered into the PRA code. For example, if the basic-event model is the exponential distribution with parameter λ, data about the distribution of this parameter are entered into the code. The distributions of the parameters of all basic events are used subsequently to propagate parameter uncertainty through the PRA model.

Step 3: Define Epistemic Correlation Groups

When evaluating the PRA model to assess a risk metric or an intermediate value, such as the frequency of an accident sequence, the correlation between the estimates of the parameters of some basic events of the model needs to be taken into account. The correlation occurs because, for basic-event models employing the same parameters, the state of knowledge about these parameters is the same. In other words, the events are not independent but are related to each other. If the EC is ignored, the metric's mean value and uncertainty may be underestimated. Appendix A to this section discusses the fundamental principles of the EC.

The first step in accounting for the EC between basic events is identifying correlated events, and the outcome is the identification of several groups of correlated basic events. Each group contains basic events that are correlated with each other because the state-of-knowledge of the analysts about these events' parameters is the same.

Identifying correlated basic events principally involves determining what basic-event models share the same parameters. For example, for all components of a certain type in a nuclear power plant (NPP), if the failure rate for its failure mode is evaluated from the same data set, the basic events for these components are correlated. However, all the components of a certain

47

type in a NPP do not have to be correlated. For example, if the failure rates for subgroups are determined using different data sets, the basic events for these components are correlated within the subgroups, but not across the subgroups. Accordingly, for a particular PRA model, several different groups of correlated basic events can be defined.

The groups of correlated basic events should not be confused with groups of common-cause failures (CCFs). Although both groups express dependencies between the components of an NPP, they express different dependencies. For this reason, accounting for one type of dependency does not necessarily account for the other. Thus, accounting for CCFs in a PRA model does not necessarily account for the dependencies due to the EC. Actually, a group of correlated basic events can contain several events, including those that are modeled within a CCF group. For instance, a CCF group may contain one failure mode of all the pumps of a particular system, while a group of correlated basic events may encompass the same failure mode for all the pumps of this type within the NPP. Hence, both types of dependencies (i.e., CCF and EC) should be included in a PRA model.

Step 4 (CC II): Establish the Significance of the EC

For CCs II and III, the ASME/ANS PRA Standard requires that the mean for a risk metric is provided. This mean is calculated ensuring that the epistemic correlation is taken into account. For CC II, however, the evaluation of the mean may ignore this correlation provided it is demonstrated be not to be significant for the particular case. The interpretation of significant within this context is the contribution that the EC makes to the quantitative results of a risk metric.

Ideally, a risk metric would be evaluated taking into account the EC, and so no need would exist to establish whether the EC is significant. This preferred approach is the one required by CC III because it accounts for the EC just as any other dependency that the PRA model explicitly includes.

In some cases, the PRA model already was quantified without accounting for the EC. However, it may be possible to establish the significance of the EC without the need to explicitly account for the EC in the PRA model.

Section 2.4, "Guidelines for Addressing Parametric Uncertainty," of the Electric Power Research Institute's (EPRI) report 1016737 [EPRI, 2008] presents some brief guidelines for addressing the EC in evaluating the mean value of a risk metric and its associated parameter uncertainty. That section considers two cases for meeting the QU-A3 and QU-E3 supporting requirements: (1) the base PRA model and (2) a PRA application. For the first case (i.e., the base model), it is recommended that the EC is appropriately accounted for by setting up the groups of basic events that are correlated in the model and propagating the parameter uncertainty in the model. This approach would suffice for satisfying Capability Category III for these SRs in assessing the mean value of a risk metric and its associated parameter uncertainty. If this preferred approach cannot be completed, that section gives two separate guidelines: one for estimating the mean value of a risk metric, and one for estimating its associated parameter uncertainty.

For the second case (i.e., for a PRA application), Section 2.4 of the EPRI report proposes somewhat different approaches for evaluating the mean value of a risk metric and its associated parameter uncertainty. For assessing the mean value, it provides two guidelines. The preferred approach consists in appropriately accounting for the EC by setting up the groups of basic events that are correlated in the model and propagating the parameter uncertainty in the model.

48

The application of this guideline would yield the mean value of a risk metric. Alternatively, the cutsets of the PRA model of the application can be reviewed to establish whether the risk metric used for the application is determined by cutsets that involve basic events with epistemic correlations. If they do not, then the point estimate of the risk metric can be used instead of the mean value. However, this guideline may not be practical to implement because it would require a detailed review of cutsets that have an impact on the risk metric to determine if basic events that are correlated are present.

Section 2.4 of the EPRI report also presents two guidelines if the parameter uncertainty of a risk metric of a PRA application has to be provided for decisionmaking. The first guideline requires demonstrating that the probability distribution is not expected to significantly change (e.g., because the significant contributors for the application do not involve correlated basic events) from the base-model probability distribution. If this condition is satisfied, the base-model probability distribution is used for the application. If it is not, the application of the second guideline appropriately accounts for the EC in evaluating the parameter uncertainty of a risk metric of a PRA application by setting up the groups of basic events that are correlated in the model and propagating the parameter uncertainty in the model.

Step 4 (CC III): Set Up EC Groups in PRA Code

Each group of correlated basic events in the PRA model should be set up in a PRA computer code such that the particular code recognizes that the basic events are correlated. In this way, a single distribution is applicable to all the basic events in a correlated group. Then, when the code propagates the uncertainty, each sample from the distribution of a group of correlated basic events is used for all the basic events in the group. These values of the basic events subsequently are used in propagating parameter uncertainty through the PRA model to generate a value of the risk metric being evaluated, such as the CDF. This evaluation process is repeated for all the samples evaluated by the code.

Step 5 discusses propagating the parameter uncertainty in the PRA model. An outcome of this uncertainty evaluation is the mean of the risk metric that accounts for the epistemic correlation in addition to other characteristics of the distribution of this metric, such as the probability distribution of the risk metric.

Step 5: Propagate Parameter Uncertainty in the PRA Model

An uncertainty evaluation of the PRA model is accomplished by executing an uncertainty run of the model in the PRA computer code to obtain a risk metric, such as CDF. The PRA model earlier was set up in the code according to Steps 1 thru 4, as applicable. Provided that the PRA model was set up such that the code recognizes that some events are correlated (as described in Step 4), the code automatically accounts for the epistemic correlation when running the uncertainty evaluation unless a specific code requires additional steps.

The uncertainty can be evaluated using the Monte Carlo or Latin Hypercube Sampling (LHS) methods. The number of samples used should be such that the sampling distribution obtained converges to the true distribution of the risk metric. This can be done, for example, by calculating the standard error of the mean (SEM):

$$SEM = \frac{\sigma}{\sqrt{n}}$$

where σ is the standard deviation of the sampling distribution (i.e., the square root of the variance of this distribution) and n is the number of trials. This equation indicates that large samples produce more accurate estimates of the sampling mean than do small samples.

Using this approach, an iterative process is required in which several consecutive uncertainty runs are executed, each with an increasing number of samples. At the end of each run, the SEM is calculated. The process can be stopped when (1) increasing the number of samples does not significantly change the SEM or (2) the SEM reaches a predefined small error value.

It is advisable, though not necessary, to use the LHS method, rather than the Monte Carlo approach. The reason is that employing the LHS typically significantly reduces the time required for an evaluation because it requires fewer samples than Monte Carlo to attain similar accuracy.

The results of the uncertainty evaluation of the PRA model include the mean value and the probability distribution of the risk metric evaluated, such as the CDF. The probability distribution of the risk metric sometimes is simply characterized by its 5th and 95th percentiles, which define a 90-percent confidence interval of the risk metric.

4.2 Parameter Uncertainty in Importance Measures

In some cases, decisionmaking may not be directly based on a risk metric, such as CDF or LERF, but on other quantitative measures of risk. For example, the importance measures, such as the Fussell-Vesely (FV) and risk achievement worth (RAW), provide an indication of the importance of a structure, system, or component (SSC) to the overall risk of a nuclear power plant. Hence, it also is relevant to assess the parameter uncertainty of these measures when they are calculated. Currently, the treatment of these measures is beyond the scope of the ASME/ANS PRA Standard.

A significant example of using importance measures is associated with the new regulations in 10 CFR 50.69 that allow relaxation of some "special treatment" requirements based on assessed safety significance. Special treatment requirements are intended to ensure the proper performance of safety-related SSCs. The nuclear industry proposed a process for categorizing SSCs according to risk significance, essential to which is the use of the FV and RAW importance measures.

In calculating an importance measure, parameter uncertainty can be propagated in the same way that uncertainty is propagated when evaluating a risk metric such as the CDF or LERF. Such a capability recently was added to the latest version of the computer code SAPHIRE, being developed for the NRC by the Idaho National Laboratory.

4.3 Input to the Decisionmaker

This section provides the guidance on treating parameter uncertainty when comparing PRA results with acceptance guidelines. The relevant input to the decisionmaker consists of three elements: (1) an estimate of the relevant risk metric(s), usually expressed as the mean value(s); (2) the probability distribution(s) of these risk metric(s); and (3) the acceptance guidelines to be used for the particular application. Using these elements, the decisionmaker can understand if and how these guidelines are satisfied, and the parameter uncertainty associated with this application.

The parameter uncertainties are characterized by the probability distributions obtained for the parameter(s) of the basic event models, and the parameter uncertainties are propagated to obtain the uncertainties for the intermediate and final results of the analysis. These results, along with their uncertainties, are then compared to the acceptance guidelines. Because most acceptance guidelines currently are explicitly or implicitly couched in terms of the mean values of the risk metrics, robust mean values usually are the key results from the risk analysis.

Several different approaches previously were advocated for comparing PRA results with acceptance guidelines. SECY-97-221, "Acceptance Guidelines and Consensus Standards for Use in Risk-Informed Regulation" [NRC, 1997a], discusses the comparison of these results with numerical acceptance guidelines and alternative approaches for making the comparisons. SECY-97-287, "Final Regulatory Guidance on Risk-Informed Regulation: Policy Issues" [NRC, 1997b] summarizes the various comparative approaches. These approaches are not restricted to parameter uncertainties but apply to any uncertainties characterized in terms of a probability distribution on the value of a risk metric wherein the probability associated with a particular value represents a measure of the analyst's degree of belief that the value is a bound on the true value. SECY-97-221 identified three possible approaches for comparing such distributions with acceptance guidelines.

The first approach involves comparing the mean values of the metrics (and possibly their increments) with the acceptance guidelines. Using mean values conceptually is simple and consistent with classical decisionmaking. Moreover, an evaluation of the mean value incorporates a consideration of those uncertainties that the model explicitly captures. However, as pointed out in SECY-97-287, the mean, as with any other single estimate derived from a distribution, is a summary measure that does not fully account for the information content of the probability distribution.

The second approach involves using percentile measures. Assuming the uncertainty is characterized in terms of a probability distribution of the numerical PRA result, then an approach would be to overlay the distribution on the value associated with the acceptance guidelines (which could be expressed as a single value or, in theory, could also be in terms of a probability distribution) and to determine at what level of confidence the guidelines are met. This would require a policy decision about an acceptable level of confidence. Historically, assurance levels of 0.95 are typically seen as being characteristic of acceptability.

Such an approach is intuitively appealing, but as pointed out in SECY-97-221, several concerns exist. First, the forms of the distributions for characterizing the parameter uncertainties are arbitrary. In particular, when deciding on an appropriate distribution, most of the focus is on the central 90 percent of the distribution and the tails of the distributions usually receive little attention. Therefore, comparison to high-percentile values involving the tails of the distribution may be overly conservative or may give a false sense of assurance. In addition, the question of the acceptable level of confidence is not just about how much of the distribution lies above the acceptable value but also how it is distributed. Finally, because many of the existing NRC guidelines, such as the Commission's Safety Goals and their subsidiary objectives, were meant to be compared with mean values, this option would require a reevaluation of the guidelines themselves.

A third alternative is one where one guideline is defined for comparison to the mean value of the distribution characterizing the uncertainty and a companion guideline is defined for comparison with, for example, the 95th percentile. This approach treats the acceptance guidelines not as a simple go/no-go limit but rather as a tolerance band. It would necessitate a change in policy to determine the form of the acceptance guidelines and, in particular, to establish the upper (percentile) guidelines. Again, the mean, and especially the higher percentiles, are sensitive to fluctuations in the tails of the distributions.

As SECY-97-287 points out, from a theoretical standpoint, no clear advantage exists to choosing any one of these approaches over the others. They are all subject to the criticisms that the complete distribution is not fully used, that the form of the distributions for characterizing the input uncertainties is arbitrary to some extent, and that both the mean and the higher percentiles are sensitive to changes in the tails of the distributions.

SECY-97-287 recommends that parametric uncertainty (and any explicit model uncertainties) in the assessment be addressed using mean values to compare to acceptance guidelines. Moreover, the SECY recommends using sensitivity studies to evaluate the impact of using alternate models for the principal implicit model uncertainties. To address incompleteness, the SECY advocates employing quantitative or qualitative analyses as necessary and appropriate to the decision and to the acceptance guidelines. The mean value (or other appropriate point estimate if it is arguably close enough to the mean value) is appropriate for comparing with the acceptance guidelines. This approach has the major advantage that it is consistent with the state-of-the-art because current acceptance guidelines (i.e., the Commission's Safety Goals and subsidiary objectives) were meant to be compared with mean values. The SECY also points out that for the distributions generated in typical PRAs, the mean values typically corresponded to the region of the 70th to 80th percentiles. Coupled with a sensitivity analysis focused on the most important contributors to uncertainty, these mean values can be used for effective decisionmaking.

The form of the acceptance guidelines also will play a role in determining the appropriate uncertainty comparison. For example, the acceptance guidelines for PRA results in Regulatory Guide 1.174 [NRC, 2002] require comparison against the risk metrics of CDF and LERF and their increments. In this example, the means of the risk metrics and the means of their increments need to be established for comparison with the figure of acceptable values.

In summary, the present recommended practice is to use the mean of the risk metric to compare to the relevant acceptance guidelines; additionally, the probability distribution of the risk metric also may be presented to the decisionmaker. The actual information presented depends on the ASME/ANS PRA Standard CC of the PRA. As previously discussed in the guidance on meeting the PRA Standard SRs in this section, in SR QU-A3 for CCs II and III, the Standard requires providing the mean for a risk metric. For CC I, only the point estimate is required and used in lieu of the mean when comparing with the relevant acceptance guidelines. The difference between CCs II and III is that in the latter, the mean is calculated ensuring that the epistemic correlation is taken into account; for CC II, the evaluation of the mean may ignore this correlation provided it is demonstrated that this correlation is not significant for the particular case.

Similarly, the information about parameter uncertainty that is provided depends on the PRA CC. For CC I, the SR QU-E3 only calls for an estimate of the uncertainty interval and its basis. For CCs II and III, the Standard is interpreted as requiring that the probability distribution for a risk metric is obtained by propagating parameter uncertainty through the PRA model using a Monte Carlo approach or similar means. Again, the difference between CCs II and III is that in the latter, the probability distribution is evaluated ensuring that the epistemic correlation is taken into account; for CC II, the calculation of the probability distribution may ignore this correlation provided this correlation is demonstrated to be not significant for the particular case being assessed.

Section 7 further discusses the use of the mean and probability distribution information when comparing to acceptance guidelines.

APPENDIX 4-A: THE EPISTEMIC CORRELATION

4-A.1 Definition of the Epistemic Correlation

As mentioned in Section 3, many of the acceptance guidelines used in regulatory decisionmaking (e.g., the acceptance guidelines of Regulatory Guide 1.174 [NRC, 2002]) are defined such that the appropriate measure for comparison is the mean value of the uncertainty distribution of the relevant risk metric. Because ignoring the epistemic correlation (EC) may yield an underestimate of the mean value (and probability distribution) of a risk metric, care should be exercised to account for the EC so that the resulting metric(s) can be compared meaningfully with such guidelines.

The EC stems from the fact that, for identical or similar components in a given nuclear power plant (NPP), the state of knowledge about their failure parameters is the same, as explained below. Apostolakis and Kaplan [Apost., 1981] described this correlation, and parts of this discussion are based on their paper.

An analyst's state of knowledge about the possible values of a parameter θ is expressed in terms of a probability distribution, such as $h(\theta)$, when using Bayesian updating or expert judgment. For a particular basic event i, its associated probability distribution is denoted here by $h_i(\theta_i)$.

It is common practice to assign the same value to the parameters of basic events of identical or similar components. Therefore, for example, the probability of failure of a class of identical motor-operated valves (MOVs) to open is considered the same.

In evaluating the Probabilistic Risk Assessment (PRA) model of a specific NPP, suppose that θ_1 and θ_2 represent the parameters of two physically distinct but identical MOVs. Because this discussion assumes that all such MOVs have the same parameter, it is necessary to set $\theta_1 = \theta_2$. Moreover, because the analyst's state of knowledge is the same for the two valves, it follows that

$$h_1(\theta_1) = h_2(\theta_2)$$ Equation 4-A-1

Thus, in evaluating the PRA model to obtain attributes associated with the parameter uncertainty of a risk metric of a NPP (such as the variance of the distribution of this metric), h_1 and h_2 must be regarded as being equal distributions and treated as completely dependent distributions. The variance is a measure of spread (i.e., width) of the distribution of the risk metric.

The EC applies to distributions of parameters, such as $h(\theta)$, and also to other kinds of basic events, such as common cause failures and human errors. Forgetting to return valves to their normal position after testing is an example of similar-type human errors.

4-A.2 Effect of the EC on a Risk Metric and Associated Uncertainty

The mean of a minimal cutset (MCS)[10] containing basic events that are correlated is underestimated because

$$E(X^n) > E^n(X) \qquad \text{Equation 4-A-2}$$

where X is a random variable corresponding to a basic event that is correlated with other basic events in the MCS, $E(X^n)$ is the expected value of the random variable X elevated to the nth power and $E^n(X)$ is the nth power of the expected value of X.

To illustrate this underestimation for n = 2, consider the simple case where two MOVs are in parallel, represented by variables X_1 and X_2 that are correlated, and system failure occurs when both fail to open. The correct equation is

$$T = X^2 \qquad \text{Equation 4-A-3}$$

where T represents system failure. This equation expresses the fact that the failure probabilities of the two MOVs are identical (i.e., the distributions of the failure probabilities express the same state of knowledge).

If X_1 and X_2 are incorrectly considered to be independent, the equation used for system failure would be

$$T = X_1 X_2 \qquad \text{Equation 4-A-4}$$

This equation underestimates the mean of T, as can be seen by taking the expected value in Equations 4-A-3 and 4-A-4. Thus, using Equation 4-A-3,

$$E(T) = E(X^2) = E^2(X) + \sigma^2_X \qquad \text{Equation 4-A-5}$$

where $E^2(X)$ is the second power of the expected value of X and σ^2_X is the variance of X.

Using Equation 4-A-4,

$$E(T) = E(X_1 X_2) = E^2(X) \qquad \text{Equation 4-A-6}$$

Comparing Equations 4-A-5 and 4-A-6 demonstrates that the mean value of the system failure (i.e., the expected value of this failure E(T)) is underestimated when the EC is ignored.

The underestimation of the mean of an MCS that contains correlated basic events is particularly significant when

$$E(X^n) \gg E^n(X) \qquad \text{Equation 4-A-7}$$

This condition occurs when an MCS contains more than two basic events that are correlated or when the uncertainty (i.e., spread) of the distribution of the correlated basic events in an MCS is large.

---

[10] An MCS is a minimal set of basic events that causes an undesired event, such as core damage.

This example of two MOVs in parallel also serves to illustrate the potential underestimation of the uncertainty as expressed by the variance of the distribution of system failure. Considering that the distributions of the failure probabilities of the two MOVs express the same state of knowledge, the correct equation is

$$\sigma^2_T = E(X^4) - E^2(X^2)$$
Equation 4-A-8

where σ^2_T is the variance of T.

If X_1 and X_2 incorrectly are considered to be independent, the equation used for the variance of the distribution of system failure would be

$$\sigma^2_T = E^2(X^2) - E^4(X)$$
Equation 4-A-9

In typical evaluations, Equation 4-A-8 yields a greater variance (i.e., uncertainty) than Equation 4-A-9. It is important to note that the uncertainty of the distribution of system failure potentially will be underestimated even if the correlated events are not in the same MCS.

Example. The simple system containing two MOVs in parallel is evaluated using data from the paper by Apostolakis and Kaplan to illustrate the quantitative effect of employing incorrect equations for estimating the mean and uncertainty expressed in terms of the variance. This example is presented in the font Times New Roman to distinguish it from the rest of the text. The data used are the following:

$E(X) = 1.5 \times 10^{-3}$
$E(X^2) = 6.0 \times 10^{-6}$
$E(X^4) = 1.6 \times 10^{-9}$

Using these data, the variance of X is

$\sigma^2_X = E(X^2) - E^2(X) = 3.8 \times 10^{-6}$

Table 4-2 lists the correct and incorrect values of the mean and variance of system failure and the factor by which these values differ. For this simple example, the mean and variance are underestimated, respectively, by factors of about 2.7 and 50.6.

Table 4-A-1   Example of quantitative effect of failing to account for EC.

Parameter	Correct value	Incorrect value	Factor
Mean, E(T)	6.0×10^{-6} (Equation 4-A-5)	2.3×10^{-6} (Equation 4-A-6)	2.7
Variance, σ^2_T	1.6×10^{-9} (Equation 4-A-8)	3.1×10^{-11} (Equation 4-A-9)	50.6

Accordingly, failing to take into account the EC when evaluating a risk metric or an intermediate value, such as the frequency of an accident sequence, has the following potential impacts:

- For MCSs containing correlated basic events, ignoring the EC will underestimate the mean of each MCS containing such events. This point has implications in generating the MCSs from the logic model of the PRA and in using the mean values of the MCSs to estimate the mean of the intermediate results and final risk metrics, as follows:

–   Screening MCSs may incorrectly delete some of them. This may occur in two contexts. In one, because the number of MCSs in a PRA model can be extremely large, it is common to use a truncation value in solving the PRA's logic model. In this way, only MCSs above this value are obtained, while the rest are neglected. Should the mean value or other point estimate of an MCS be assessed without accounting for the EC, and this value used for comparison with the truncation value, some MCSs containing basic events that are correlated may be incorrectly eliminated when generating the cutsets.

To illustrate this concern in the first context, assume that a PRA model is evaluated using a truncation value of 1×10^{-9} per year. If the frequency of the MCSs are calculated using a point estimate that does not account for the EC, and the point-estimate frequency of each MCS containing correlated basic events is smaller than this value, a subset of them may be incorrectly discarded because the correct frequency (that accounts for the EC) of each MCS in this subset is actually larger than this value. The significance of this inappropriate elimination to the estimate of the frequency of a risk metric depends on the combination of two factors: (1) the correct frequency of the risk metric and (2) the correct frequency of the MCSs in the mentioned subset. If the point-estimate frequency of the risk metric is not too large compared to the truncation value, say 1×10^{-6} per year, and the correct frequency of the MCSs in the mentioned subset is large compared to the truncation value, say 1×10^{-7} per year, the elimination is significant. This issue may become more important for evaluating the mean and the parameter uncertainty in PRAs of future plants, whose risk metrics are expected to have even lower values than those for current plants.

In the other context, after the MCSs have been generated from the logic model of the PRA, sometimes they are screened to identify some for detailed evaluation, thus eliminating others from further consideration. For example, when carrying out a PRA application, it may be desirable to work with a subset of the MCSs that were generated when the complete PRA model was solved. This subset may be obtained by selecting the MCSs above a cutoff value. Consideration of the EC can be important when screening the MCSs using their means because some MCSs may be incorrectly deleted.

–   Underestimating the mean of the risk metric. The means of the MCSs above the truncation value are combined to estimate the mean of the risk metric, and some of these MCSs may contain basic events that are correlated. Hence, underestimating the mean of these MCSs may cause an underestimation of the mean of the risk metric.

• In evaluating the parameter uncertainty of the risk metric, ignoring the EC will underestimate the uncertainty.

The combined effect of incorrectly removing some MCSs (below the truncation value) from the quantitative evaluation and underestimating the mean of MCSs (above the truncation value) containing basic events that are correlated potentially will result in a cumulative underestimation of the mean of the risk metric or other intermediate values.

One approximate approach is proposed in NUREG/CR-4836 [SNL, 1988] for addressing the issues raised in the bullet "Screening of MCSs may incorrectly delete some MCSs," above. The practicality of this approach remains to be demonstrated for large-scale PRAs.

In summary, failing to take into account the EC when evaluating a risk metric or an intermediate value potentially might underestimate the mean and the uncertainty of the distribution of this metric. The simple example above where two MOVs are in parallel clearly reveals that the underestimation of the mean can be significant, especially if the variance of the distribution of the correlated events is large.

5. MODEL UNCERTAINTY

This section provides guidance for addressing sources of model uncertainty and related assumptions related to the base probabilistic risk assessment (PRA). The guidance focuses on sources of model uncertainty and related assumptions in the context of the requirements in the American Society of Mechanical Engineers (ASME)/American Nuclear Society (ANS) PRA Standard [ASME/ANS, 2009] and in the context of risk-informed decisionmaking. Consequently, the guidance in this section focuses on identifying and characterizing sources of model uncertainties and related assumptions in a PRA and then assessing what might be their impact on the PRA results and insights used to support risk-informed decisions. In particular, the section provides guidance on the following:

- Definitions of key sources of model uncertainty and related key assumptions.

- A process used to identify and characterize the sources of model uncertainty and related assumptions in probabilistic risk assessment (PRA) that are key to the decision under consideration.

5.1 Definitions

To provide guidance on how to address the impact of model uncertainty and related assumptions, it is necessary to understand what is meant by these terms. Table 5-1 provides the following definitions that are consistent with Regulatory Guide (RG) 1.200 [NRC, 2007a][11] and the ASME/ANS PRA Standard.

Table 5-1   Definitions for model uncertainty and related assumption.

A *source of model uncertainty* is one that is related to an issue in which no consensus approach or model exists and where the choice of approach or model is known to have an effect on the PRA (e.g., introduction of a new basic event, changes to basic event probabilities, change in success criterion, introduction of a new initiating event).

A source of model uncertainty is labeled *key* when it could impact the PRA results that are being used in a decision and, consequently, may influence the decision being made. Therefore, a key source of model uncertainty is identified in the context of an application. This impact would need to be significant enough that it changes the degree to which the risk acceptance criteria are met and, therefore, could potentially influence the decision. For example, for an application for a licensing basis change using the acceptance criteria in RG 1.174 [NRC, 2002], a source of model uncertainty or related assumption could be considered "key" if it results in uncertainty regarding whether the result lies in Region II or Region I or if it results in uncertainty regarding whether the result becomes close to the region boundary or not.

---

[11] Revision 2 to RG 1.200, which endorses ASME/ANS PRA Standard RA-Sa-2009 [ASME/ANS, 2009], is being finalized at the time of publication of this NUREG, and should be available in April 2009.

Table 5-1   Definitions for model uncertainty and related assumption.

An *assumption* is a decision or judgment that is made in the development of the PRA. An assumption is either related to a source of model uncertainty or to scope or level of detail.

An *assumption related to a model uncertainty* is made with the knowledge that a different reasonable alternative assumption exists. A *reasonable alternative assumption* is one that has broad acceptance within the technical community and for which the technical basis for consideration is at least as sound as that of the assumption being made.

An *assumption related to scope or level of detail* is one that is made for modeling convenience.

An assumption is labeled *key* when it may influence (i.e., have the potential to change) the decision being made. Therefore, a key assumption is identified in the context of an application.

Some explanation on the definition of an assumption related to a model uncertainty is required. The identification of reasonable alternative assumptions is, as will be discussed later, relevant to the choice of sensitivity analysis cases used to test the robustness of the results. Even though an alternative assumption does exist, it is possible that it may not have broad acceptance within the technical community. For an assumption to be considered as a reasonable alternative for the purposes of this section, it should have a sound technical basis. In some cases, no reasonable, alternative assumptions may exist. In these cases, the assumptions and their related models become de facto consensus models in the sense that no alternative is available.

Although the focus of this section is on the treatment of key sources of model uncertainty and related assumptions, those assumptions related to scope and level of detail also can have an effect on the results of the model and need to be considered. Section 7 discusses this further.

5.2   Overview of Process

Figure 5-1 illustrates the three-step process used to identify the sources of model uncertainty and related assumptions from the base PRA that are key to the decision under consideration.

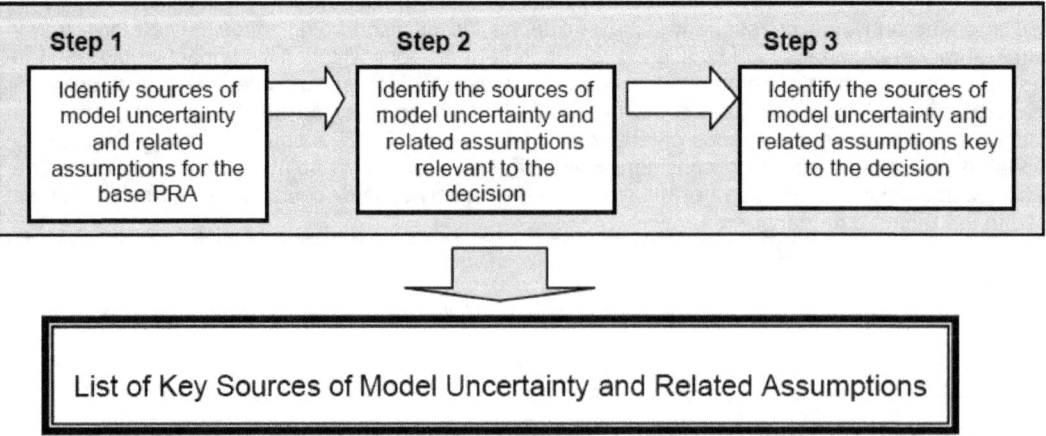

Figure 5-1   Process to develop list of key sources of uncertainty and related assumptions

Step 1: <u>Identify Sources of Model Uncertainties and Related Assumptions in the Base PRA</u> — The base PRA is reviewed to identify and characterize the sources of model uncertainty and related assumptions. Some sources may be generic, and some may be plant specific. These sources of model uncertainty and related assumptions are those that result from developing the PRA.

Step 2: <u>Identify Sources of Model Uncertainties and Related Assumptions Relevant to Decision</u> — The sources of model uncertainty and related assumptions associated with the base PRA are reviewed to identify those that are relevant to the decision under consideration. New sources of model uncertainty and related assumptions that may be introduced by the application also are identified. This identification is based on an understanding of the type of application and the associated acceptance guidelines.

Step 3: <u>Identify Sources of Model Uncertainties and Related Assumptions Key to the Decision</u> — The sources of model uncertainty and related assumptions that are relevant to the decision are reviewed to identify those that are key to the decision. This review involves performing a quantitative analysis to identify the importance of each relevant source.

These steps are described in more detail below.

5.3 Step 1: Base PRA Sources of Model Uncertainties and Related Assumptions

This step involves identifying and characterizing those sources of model uncertainty and related assumptions in the base PRA. As part of this step, guidance is provided on accomplishing QU-E1, QU-E2, QU-E4, and LE-F3 supporting requirements in the ASME/ANS PRA Standard that involves model uncertainties.

- QU-E1 – Identify sources of model uncertainty.

- QU-E2 – Identify assumptions made in the development of the PRA.

- QU-E4 – For each source of model uncertainty and related assumptions identified in QU-E1 and QU-E2, respectively, identify how the PRA is affected (e.g., introduction of a new basic event, changes to basic event probabilities, change in success criterion, introduction of a new initiating event.

- LE-F3 – Characterize the LERF sources of model uncertainty and related assumptions.

Once sources of model uncertainty and related assumptions have been identified, they may be evaluated against qualitative screening criteria to identify those sources that can be eliminated as being key sources of model uncertainty based on the existence of some accepted model approach, such as a consensus model for a particular issue. Figure 5-2 illustrates the process used to identify and characterize the sources of model uncertainty and related assumptions. This process, discussed below, involves the following:

- Identification of sources of model uncertainty and related assumptions.
- Characterization of sources of model uncertainty and related assumptions.
- Qualitative screening of model uncertainty and related assumptions.

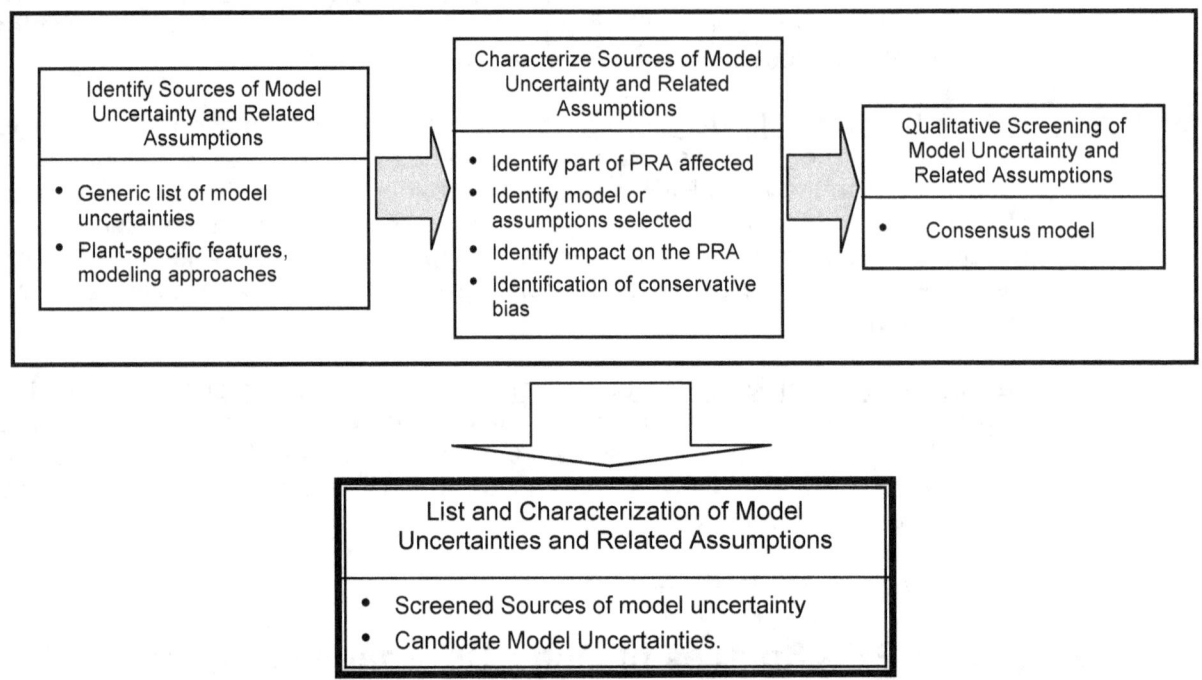

Figure 5-2   Process to identify and characterize the sources of model uncertainty and related assumptions

The process of Step 1 is consistent with the process discussed in Section 3 of "Treatment of Parameter and Model Uncertainty for Probabilistic Risk Assessments," [EPRI, 2008], hereon referred to as the Electric Power Research Institute (EPRI) report.

5.3.1 Identification

The first part of Step 1 involves identifying the sources of model uncertainty and related assumptions that are a result of developing the PRA. This identification is performed by using the structure of the PRA and, more explicitly, by examining each step of the construction of the PRA to identify if it involved a model uncertainty or related assumption as defined previously in Section 5.1. This process provides a systematic method of identifying the sources of model uncertainty and related assumptions. This process basically involves using the ASME/ANS PRA Standard as the PRA structure and reviewing each supporting requirement to determine if a model uncertainty or related assumption is involved. This identification is required in the ASME/ANS PRA Standard. For each technical element, the standard requires that the sources of model uncertainty are identified and documented.

For this process, the EPRI report provides an example of an acceptable approach. Moreover, the EPRI report provides a generic list of sources of model uncertainty and related assumptions as a result of implementing the process (see Tables A-1 and A-2 of the EPRI report). This list can serve as a starting point to identify the set of plant-specific sources of model uncertainty and related assumptions. That is, the tables in the EPRI report are a generic list. The analyst is expected to apply the process to also identify any unique plant-specific source.

5.3.2 Characterization

The second part of the Step 1 involves characterizing the identified sources of model uncertainty and related assumptions. This characterization involves understanding how the PRA can be affected by the sources of model uncertainty and related assumptions. This characterization (required by QU-E4 of the standard) involves identifying the following:

- The part of the PRA affected.

- Modeling approach used or assumption made.

- Impact on the PRA (e.g., introduction of a new basic event, changes to basic event probabilities, change in success criteria, introduction of a new initiating event).

- Identification of conservative bias.

Part of the PRA Affected

The part of the PRA affected by the source of model uncertainty or a related assumption is identified. This identification is needed because not every application involves every aspect of the PRA. Therefore, as discussed below in Step 2, if the application deals with an aspect of the PRA not affected by the source of model uncertainty, then the source is not relevant to the application.

The sources of model uncertainties could impact the PRA by affecting:

- A single basic event.
- Multiple basic events.
- The logic structure of the PRA event trees or fault trees.
- A combination of both basic events and portions of the logic structure.

Moreover, the impact on the logic structure also could occur simply in a single place or throughout the PRA in multiple places.

Three types of basic events are found in PRAs. These events represent the following:

- Occurrence of initiating events.

- States of unavailability or failure of structures, systems, and components (SSCs).

- Human failures that contribute to the failure of the systems designed to protect against the undesirable consequences should an initiating event occur.

Uncertainties and assumptions exist that can influence the frequency of initiating events, human error probabilities, and the failure probabilities of SSCs. For example, assumptions used in the human reliability analysis can affect the human error probabilities used in the model. Similarly, different approaches are available to generate loss-of-coolant accident (LOCA) frequencies. Uncertainty in deterministic analyses also can influence basic event failure rates. For example, calculations used to assess the potential for sump plugging will influence the probability assigned to that event.

Uncertainty can be associated with the necessity to include specific initiating events, the necessity to include specific failure modes of SSCs, the means to address the consequences of SSC failure (e.g., digital instrumentation and control systems), the means to assess success criteria, the necessity to develop; support systems for specific missions of front-line systems, the necessity to model failures of complex pieces of equipment (e.g., RCP seals), and other issues. The way in which these uncertainties are addressed can have an impact on accident sequence definitions or system fault tree structures.

For example, an uncertainty associated with the establishment of the success criterion for a specific system can result in an uncertainty as to whether one or two pumps are required for a particular scenario. This uncertainty would be reflected in the choice of the top gate in the fault tree for that system. On the other hand, those model uncertainties that are associated with choosing the model to estimate the probabilities of the basic events do not alter the structure of the model.

Modeling Approach Used or Assumption Made

At this point in the process, the modeling approach or related assumption selected for each source of model uncertainty is identified. Different models or different assumptions are available to address the identified source of uncertainty. This identification is important because, depending on the approach or assumption selected, it will determine how the PRA is affected.

For example, different models have been proposed for modeling reactor coolant pump (RCP) seal failures on loss of cooling, different thermo-hydraulic computer codes have been used to derive success criteria, different human reliability analysis (HRA) models are used for deriving human error probabilities (HEPs), and different assumptions can be made with regard to equipment performance under adverse conditions. For all these examples, the analyst's choice of model or assumption will affect the PRA. This effect may introduce, for example, new initiating events or accident sequences, or it could affect the computational results. Some examples are provided below.

Impact on the PRA

At this point, for each source identified, its potential impact on the PRA is determined. How the PRA may be impacted was identified above in characterizing the sources. Determining a source's potential impact involves identifying how the PRA results would change if an alternate model were selected. The following is a list of some examples:

- An alternate computational model may produce different initiating event frequencies, SSC failure probabilities, or unavailabilities.

- An alternate HRA model may produce different HEPs or introduce new human failure events.

- An alternate assumption regarding phenomenological effects on SSC performance can impact the credit taken for SSCs for some accident sequences.

- An alternate success criterion may lead to a redefinition of a system fault tree logic.

- An alternate screening criterion may lead to adding or deleting initiating events, accident sequences, or basic events.

- An alternate assumption that changes the credited front-line or support systems included in the model may change the incremental significance of sequences.

- An alternate seal LOCA model can result in a different event tree structure, different LOCA sizes, and different probabilities.

Identification of Conservatism Bias

An important aspect of characterizing a source of model uncertainty and related assumption is understanding whether the model chosen or assumption adopted is conservative. This understanding is necessary because for some applications, the use of conservative assumptions in one area can mask the significance of another part of the risk picture, which might be the part that is needed for an application. This is particularly true for applications that involve risk categorization or ranking. Thus, if a source of model uncertainty or related assumption is identified as resulting in a conservative bias, it is necessary to assess the impact of the conservatism on the PRA.

"Conservative," as used in this report, implies that adopting the assumption would lead to a higher risk estimate than if other reasonable assumptions were adopted. In particular, it is important to know whether the assumption is such that a more realistic assumption would result in a lower risk. A de facto consensus of acceptance may exist when certain conservative U.S. Nuclear Regulatory Commission (NRC) licensing criteria are used as the basis to model certain issues. An example of such a conservative criterion is the 2- to 4-hour coping time for battery depletion during a loss-of-alternating-current (ac) power event because station batteries are expected to be available for several more hours if loads were to be successfully shed. A model reflective of the licensee's licensing basis is generally perceived as incorporating the conservative attributes of the deterministic licensing criteria.

For this process, the EPRI report provides an acceptable approach. Moreover, the EPRI report provides a generic list of sources of model uncertainty and related assumptions as a result of implementing the process. As discussed in Section 5.3.1, for each source listed, the impact on the PRA also is provided.

5.3.3 Qualitative Screening

The last part of Step 1 involves a qualitative screening of the sources of model uncertainties and related assumptions. The purpose of this screening is to identify those sources that do not warrant further consideration as potential key sources of uncertainty and related assumptions. That is, sources of uncertainty or related assumptions may exist that, for qualitative reasons, do not need to be considered in the decisionmaking process. This qualitative screening involves determining whether a consensus model has been used to evaluate an identified model uncertainty.

Adopting a consensus model will remove an issue as a source of model uncertainty to be addressed in the decisionmaking. With the use of a consensus model, an alternative hypothesis need no longer be explored. It should be noted that adoption of a consensus model does not mean that no uncertainty is associated with its use. However, this uncertainty generally would be manifested as an uncertainty on the results used to generate the probability of the basic event(s) to which the model is applied. This uncertainty would be treated in the PRA quantification as a parameter uncertainty. What is removed by adopting a consensus model is the need to consider other models as alternatives.

The models adopted for the sources identified are reviewed to identify those that meet the definition of a consensus model and, consequently, can be screened from further consideration. The definition of a consensus model (as defined in Section 3.3.3) is as follows:

> ***Consensus model:*** In the most general sense, as a model that has a publicly available published basis and has been peer reviewed and widely adopted by an appropriate stakeholder group. In addition, widely accepted PRA practices may be regarded as consensus models. Examples of the latter include the use of the constant probability of failure on demand model for standby components and the Poisson model for initiating events. For risk-informed regulatory decisions, the consensus model approach is one that the NRC has utilized or accepted for the specific risk-informed application for which it is proposed.

It is important to note that the definition given here ties the consensus model definition to a specific application. This relationship is because models have limitations that may be acceptable for some uses and not for others.

Examples of consensus models include the following:

- Poisson model for initiating events.
- Bayesian analysis.
- Westinghouse and Combustion Engineering RCP seal LOCA models.

5.4 Step 2: Relevant Sources of Model Uncertainties and Related Assumptions

This step involves identifying those sources of model uncertainty and related assumptions that are relevant to an application. In Step 1, the sources of model uncertainty and related assumptions associated with the base PRA were identified and characterized. However, in the context of a decision, not all sources of model uncertainty may be relevant and, therefore, will not be key to the decision. Consequently, the sources of model uncertainty and related assumptions that are not relevant can be screened from further consideration.

A source of model uncertainty or related assumption identified in Step 1 is only relevant to an application if it is related to a part of the PRA used to generate the results needed to support the application. Therefore, if a source of model uncertainty or related assumption does not impact the part of the PRA relevant to the application, then it also is not relevant to the application. In addition, if the base PRA is modified to generate the results needed for the application, these modifications themselves can introduce new sources of model uncertainty or assumptions. The process used to identify those sources of uncertainty relevant to the application involves the following:

- Understanding the way in which the PRA is used to support the application.

- Identifying the sources of model uncertainty in the base PRA relevant to the PRA results needed for the application.

- Identifying and characterizing relevant sources of model uncertainty associated with the changes to the PRA.

5.4.1 Understanding the PRA Used to Support the Application

The first part of Step 2 involves understanding the way in which the base PRA model is to be used to provide the results needed to support the application. This understanding involves (a) identifying what results are needed to support the decision under consideration and (b) establishing the cause-effect relationship between the decision under consideration and the PRA model.

In identifying what results are needed, for an application, the decision under consideration will be influenced by the extent to which the acceptance guidelines are impacted. Consequently, the acceptance guidelines will dictate what results from the PRA are required. Some guidelines are given in terms of a single metric (e.g., the core damage frequency [CDF] associated with a class of accidents). Others, such as those in RG 1.174, are two dimensional and involve an evaluation of the change on CDF and large early release frequency (LERF) as well as the total CDF and LERF.

To establish the cause-effect relationship, it is necessary to understand which parts of the base PRA model need to be used to provide the results needed to support the application. Applications may involve hardware or operational changes. For some applications, only a portion of the PRA may be required to assess the issue being addressed in the application while for others the entire PRA may be required. Some examples are given in the next section. In either case, the base PRA models that are used will need to be modified to perform an assessment of the change in risk given a plant change. This change may take the form of changes to parameters of the model, an addition of one or more basic events, more significant changes to the logic structure, or a combination of the above. Some examples are given in the next section.

Tables A-3 and A-4 in the EPRI report provide additional information that could be useful in searching for application of specific sources of model uncertainty and related assumptions that help to ensure the PRA is comprehensively evaluated for relevance to the application.

5.4.2 Identifying Sources of Model Uncertainty in Base PRA Relevant to the Application

The next part of Step 2 involves screening those sources of model uncertainty and related assumptions in the base PRA that do not have the potential to be relevant to the decision under consideration. If a source of uncertainty is not related to either the parts of the PRA used to generate the results or the significant contributors to the results, then the source is not relevant to the decision.

5.4.2.1 Screening Based on Relevance to Parts of PRA Needed

Some sources of uncertainty may be screened because they are only relevant to parts of the PRA that are not exercised by the application. For example, if the application is only concerned with the LOCA accident sequences, then those sources of model uncertainty and related assumptions impacting the other sequences would not be considered relevant to the application.

For some applications (e.g., a simple allowed outage time [AOT] extension), the complete PRA may not need to be exercised (e.g., to assess the change in risk resulting from the AOT extension). Only those sources of uncertainty that affect the parts of the base PRA needed to support the application need to be retained for further evaluation. For example, when the application addresses an AOT extension for a diesel generator, only those parts of the PRA that involve the diesel generator need be exercised, namely those sequences that involve a loss of offsite power (LOOP). Thus, only uncertainties that affect these LOOP sequences would have to be considered. However, it should be noted that for an application such as this that uses the RG 1.174 acceptance guidelines, the total CDF also may need to be determined depending on the value of ΔCDF. In these cases, the uncertainties in the complete base PRA would need to be retained for further consideration.

On the other hand, an application such as the implementation of 10 CFR §50.69, "Risk-Informed Categorization and Treatment of Structures, Systems, and Components for Nuclear Power Reactors," requires the categorization of SSCs into low- and high-safety significance. Because these are relative measures, such an assessment would involve the complete PRA. It should be noted that Regulatory Guide 1.201, *"Guidelines for Categorizing Structures, Systems, and Components in Nuclear Power Plants According to Their Safety Significance,"* [NRC, 2006a] and NEI 00-04, *"10 CFR 50.69 SSC Categorization Guideline,"* [NEI, 2005b] address treatment of uncertainties in the categorization of SSCs. The guidelines provided in this NUREG do not supersede the approach in 10 CFR §50.69.

5.4.2.2 Screening Based on Relevance to Results

Other sources of uncertainty may be screened because they do not have an impact on the results used to support the application. The first step is to identify the significant contributors to the results. This may be done in a number of ways. One approach is to rank the basic events that drive the results using importance measures, such as Risk Achievement Worth and Fussell-Vesely. For some of the basic events this leads directly to the identification of a source of model uncertainty. This is the case when that uncertainty is associated with evaluating the basic event probability. In these cases, the Risk Achievement Worth may be useful in quantitatively screening the source of uncertainty as discussed in Section 5.5.

For other sources of uncertainty, the search is more subtle, and involves understanding the reasons why a particular basic event is important. Examining the cut-sets generating the results and the associated accident scenarios is one means of developing such an understanding. The reason for an event being important could be an assumption that has been made in developing the accident sequence to which the event is a contributor. For example, an assumption that the preferred system to achieve a function is not available for a specific initiating event will result in the secondary system becoming relatively more important for that initiating event.

Care should be taken, however, to identify those assumptions that result in some potential contributors being less important than would have been the case under an alternate assumption. This issue may be of more concern when the results that are significant to the application are related to scenarios or sequences that are typically not significant contributors to CDF and or LERF.

This type of screening requires an in-depth understanding of the assumptions underlying the structure of the logic model. Absent that understanding, it is more prudent to address all sources of model uncertainty that have not been screened in accordance with Section 5.5.

> Table A-4 of the EPRI report provides additional information on sources of uncertainty that may have been screened based on an assumption that the associated scenarios are not significant.

5.4.3 Identifying and Characterizing Relevant Sources of Uncertainty Associated with Model Changes

The last part of Step 2 involves identifying any new sources of model uncertainty that may be introduced as a result of the application. As discussed above, when modifications are made to the PRA to represent the effect of a potential change or to investigate a different design option, for example, the modifications themselves may introduce new sources of model uncertainty or related assumptions. These new sources of model uncertainty and related assumptions need to be identified. The process used in Step 1 to identify sources of model uncertainties in the base PRA is repeated for the modifications to the PRA. Specifically, the PRA modifications are reviewed against the applicable ASME/ANS PRA Standard supporting requirements to determine if a model uncertainty or related assumption is involved. For example, to assess the impact of relaxing the special treatment requirements for low-risk significant SSCs, it is necessary to model the impact on the SSC reliability. No accepted model exists for this impact.

5.5 Step 3: Identification of Key Sources of Model Uncertainty and Related Assumptions

This step involves identifying those sources of model uncertainty and related assumptions that are key to the application. Although a source may be relevant, its actual impact may not be significant enough to influence the decision under consideration. Only the relevant sources of uncertainties and related assumptions that have the potential to impact the decision are considered key.

The input to this step is a list of the relevant sources of model uncertainties identified in Step 2. These sources of model uncertainties or related assumptions may now be quantitatively assessed to identify those that have the potential to impact the results of the PRA sufficiently to impact a decision regarding the application. This determination is made by performing sensitivity analyses to determine the importance of the source of model uncertainty or a related assumption to the acceptance criteria. Those determined to be important are key sources of model uncertainty or related assumptions.

Figure 5.3 illustrates the general process.

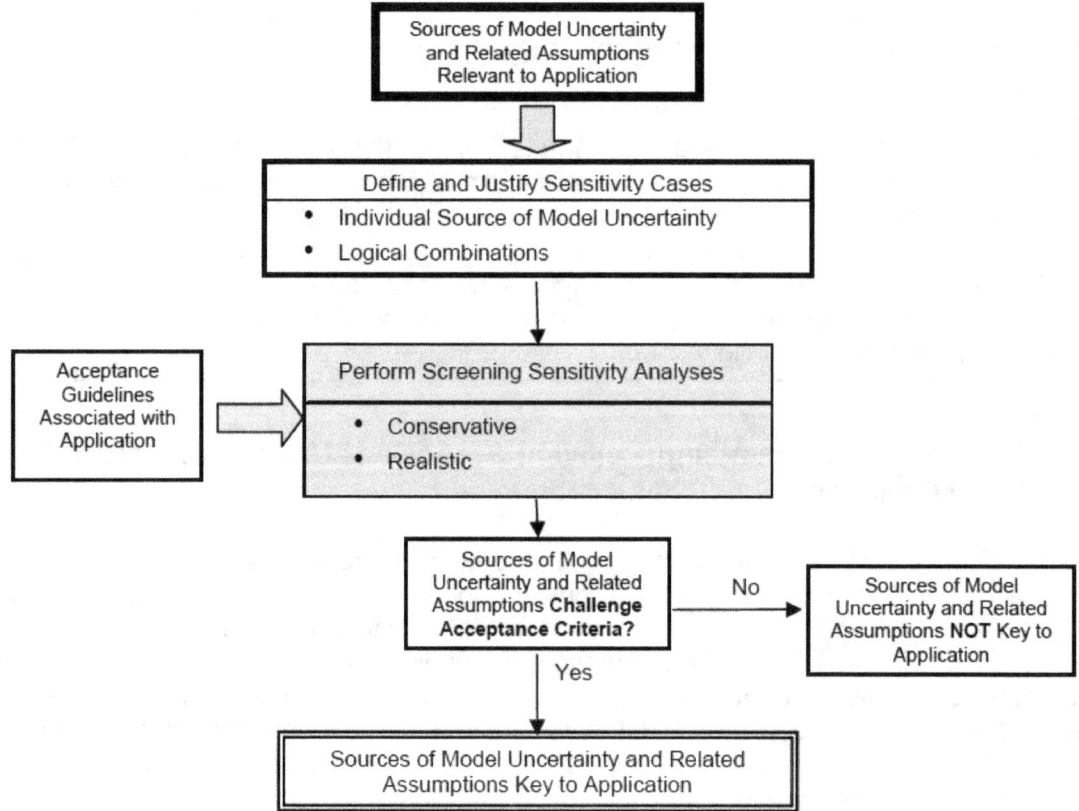

Figure 5-3   Process to identify key sources of model uncertainty and related assumptions

The process for identifying the key sources is, in principle, straightforward; however, it is dependent on the nature of the uncertainty being assessed and on the nature of the acceptance guidelines which adds some complexity. The process involves the following:

- Defining and justifying the sensitivity analyses.
- Performing the sensitivity analyses.

The sensitivity analysis determines whether the acceptance guidelines (used in the decisionmaking) are challenged. The acceptance guidelines involve either a single metric or multiple metrics. For each type of acceptance guidelines (one or two metrics), a sensitivity analysis is performed to screen those not key to the application. Moreover, for each case, the following different options can be employed in performing the sensitivity analysis:

- Only perform a conservative screening sensitivity analysis (if additional realistic sensitivity analysis is not needed).

- Initially perform a conservative screening sensitivity analysis and then perform a more realistic sensitivity analysis.

- Perform a realistic sensitivity analysis (if a conservative screening sensitivity analysis would not be useful).

Sections 5.5.2 and 5.5.3 provide guidance to address both conservative and realistic screening analyses.

5.5.1 Define and Justify Sensitivity Analysis

The first part of Step 3 involves defining an acceptable realistic sensitivity analysis. A realistic analysis involves developing reasonable alternatives or hypotheses associated with the model uncertainties relevant to the application. An alternative or hypothesis is considered to be reasonable when it has broad acceptance in the technical community and a sound technical basis.

The set of sensitivity analyses needed to obtain a reasonable understanding of the impact of the source of model uncertainty or related assumption is dependent on the particular source of model uncertainty or related assumption. To develop the alternatives, an in-depth understanding of the issues associated with the source of model uncertainty or related assumption is needed. What is known about the issue itself will essentially dictate possible alternatives to be explored.

One example of previous experience that would provide reasonable alternatives would be to identify variations in the way a particular source of model uncertainty or related assumption has been addressed in other PRA analyses, both for base PRA evaluations and for related sensitivity analyses and generally has been accepted as reasonable in the literature. One example of a common issue that has generally been accepted as reasonable in the literature is the reliability of equipment operating after loss of room cooling for which no specific calculations exist. An accepted conservative model assumption is to assign a probability of failure of 1.0. On the other hand, it may be worthwhile to explore whether the issue of room cooling is relevant to the decision by performing a sensitivity analysis under the assumption that room cooling is not needed.

Another example involves varying a parameter value to address a model uncertainty to derive a range of reasonable values. For example, consider the issue of direct current (dc) battery life. If a conservative licensing-basis model has been used, consider increasing battery life by 50 percent to represent the potential to extend battery life through operator actions. Alternatively, if a reasonable life-extension model that credits load shedding has been used, it is assumed with such a model the load-shedding procedures are performed successfully. To test the potential uncertainty impact of this assumption, one can exercise the model with a 50-percent reduction in battery life to reflect the possibility that equipment operators fail to successfully perform all tasks under stressful conditions.

It is common to use factors of, for example, 2, 5, or 10 on specific parameter values or groups of parameter values as sensitivity analyses. For these sensitivity analyses to be meaningful, it is necessary to have a justification for the factors based on an understanding of the issue that results in the uncertainty. An alternative approach to justifying the factor is to implement a monitoring program that would verify that the assumed factor is appropriate.

> For this step of the process, Appendix A of the EPRI report provides examples of alternate assumptions and methods for a number of sources of model uncertainty.

As noted earlier, for an application, the acceptance guidelines involve either a single metric or two metrics. Consequently, guidance is needed for both cases. In performing the sensitivity analysis, the sources of model uncertainty or related assumptions are linked to:

- A single basic event.
- Multiple basis events.
- The logic structure of the PRA.
- Logical combinations.

The sensitivity analyses are performed for each of the above for applications involving either a single metric or two metrics.

5.5.2 Case 1: Applications Involving a Single-Metric Acceptance Guideline

Guidance is provided for performing conservative screening or a realistic assessment for each of the four classes for applications involving a single-metric acceptance guideline. For each class, guidance is provided for both a conservative option and a realistic option.

The concept of an acceptable change in risk needs to be defined within the context of the application for which the licensee intends to use the PRA, which is the purpose of Step 2. However, it most likely would be defined in terms of a maximum acceptable risk metric, such as CDF, incremental CDF deficit, or incremental core damage probability.

5.5.2.1 Case 1a: Sources Linked to a Single Basic Event

The sources of model uncertainty and related assumptions identified in Step 2 are reviewed to determine those that are relevant only to a single basic event. For each identified source of uncertainty, an importance analysis is performed. An example is the use of an alternate method to generate a large LOCA frequency.

Conservative Screening Option

The sources of model uncertainty and related assumptions identified in Step 2 are reviewed to determine those that are relevant only to a single basic event. For each identified source of uncertainty, an importance analysis can be used for quantitative screening.

An approach to determining the importance of the source of model uncertainty or related assumptions is to calculate a maximum acceptable risk achievement worth (RAW) (denoted as RAW_{max}) associated with the metric of interest, such as maximum allowable CDF or LERF. The RAW for each relevant basic event can be compared to the RAW_{max} associated with the maximum acceptable CDF. For basic events with a RAW less than RAW_{max}, the associated model uncertainty or related assumption issue cannot by itself be key because it is not mathematically possible for the impact of that issue to cause the result of the PRA to be greater than the maximum acceptable metric. However, if it is identified as a member of a logical combination, it needs to be reconsidered.

For the jth basic event, the definition of RAW is

$$RAW_{j,base} = \frac{CDF^+_{j,base}}{CDF_{base}}$$
Equation 5-1

Where:

RAW_j is the value of RAW for basic event *j* as calculated in the base PRA

CDF_{base} is the value of the CDF mean estimate in the base PRA

$CDF^+_{j, base}$ is the base PRA CDF mean estimate with the basic event *j* set to 1.

Thus, given that the acceptance criterion defines the maximum acceptable CDF (denoted as CDF^+), Equation 5-1 can be solved directly for the "maximum acceptable" RAW, RAW_{max}, that any basic event can have without constituting a potential key uncertainty. CDF^+, the metric of interest in this case, is substituted into Equation 5-1 in place of $CDF^+_{j, base}$ and the equation is solved for RAW_{max}.

To illustrate the concept of a maximum acceptable RAW, suppose that for a particular base PRA the CDF is $3.0x10^{-5}$/year (yr). Suppose further that the maximum acceptable CDF for a particular application of the base PRA is $5.0x10^{-5}$/yr. Hence, from Equation 5-1

$$RAW_{max} = \frac{CDF^+}{CDF_{base}}$$
Equation 5-2

therefore,

$$RAW_{max} = \frac{5.0x10^{-5}}{3.0x10^{-5}}$$

$$RAW_{max} = 1.7$$

For this example, any source of model uncertainty or a related assumption linked to a basic event from the base PRA that has a RAW greater than 1.7 would have the mathematical potential to be a key source of model uncertainty or a related assumption. Such sources of model uncertainty or related assumptions should be assessed with a reasonable and realistic sensitivity analysis to determine whether or not they constitute a key source of model uncertainty or a related assumption. Expressed more generally, if:

$$RAW_{j,base} \leq RAW_{max}$$
Equation 5.3

is a true expression for the jth basic event, then the source of model uncertainty or a related assumption linked to the jth basic event is screened from further consideration. Otherwise, it is a potential key uncertainty and should be assessed with a realistic sensitivity analysis. It should be remembered that the criteria or guidelines are typically not "hard" criteria, and their application has some flexibility. However, given that the RAW measure is an extreme sensitivity measure in that it takes the failure probability to 1, this screening approach is acceptable.

It should be noted that be emphasized that the result in Equation 5-3 is relevant only for those sources of model uncertainty and related assumptions identified in Step 1 that are linked to a particular basic event.

Realistic Sensitivity Assessment Option

In performing the conservative screening, the value of a basic event is set to 1. However, a value of 1 is assuming that the *failure always occurs*. This assumption is generally not considered to be a realistic evaluation of the uncertainty associated with any particular uncertainty issue. The basic event value that would achieve a false relationship in Equation 5-3 needs to be estimated. If the value of the basic event that results in a false expression for Equation 5-3 is reasonable, then the issue is a key uncertainty. To make this determination, a reasonable alternative model for determining the basic event probability or frequency generally needs to be selected and exercised to determine the actual change in the risk metric. However, in some cases, the required basic event value may be exceedingly high that a sound argument can be made that other reasonable models would not result in this value. If the basic event value necessary to achieve a false relationship in Equation 5-3 is unreasonable, then it is not reasonable to expect that the source of model uncertainty or a related assumption could result in an undesirable risk profile. Hence, it is not a key uncertainty.

5.5.2.2 Case 1b: Sources Linked to Multiple Basic Events

The sources of model uncertainty and related assumptions identified in Step 2 are reviewed to determine those that are relevant only to multiple basic events. For each identified source of uncertainty, an importance analysis is performed. An example is an assumption that affects the quantification of a particular failure mode of several redundant components (e.g., stuck-open safety relief valve).

Conservative Screening Option

The RAW importance measures for several basic events cannot be used collectively to assess the combined impact of the sources of model uncertainties or related assumptions associated with the group of basic events. However, the concept of setting all relevant basic events to 1 simultaneously and then reevaluating the PRA yields the same perspective for a group of basic events as does the RAW importance measure analysis for an individual basic event. Hence, by setting all basic events relevant to a source of uncertainty to 1 to calculate a CDF^+_{lj} where l represents the set of basic events relevant to the lth source of model uncertainty or related assumption, the following equation can be evaluated[12]:

$$CDF^+_{l,base} \leq CDF^+ \qquad\qquad \text{Equation 5-4}$$

If the relationship in Equation 5-4 is true, then as long as it is not a member of a logical combination, the source of uncertainty is not a key uncertainty. No mathematical possibility exists that any quantification of the values of the basic events linked to that source of uncertainty could achieve an unacceptably high CDF. Otherwise, the source of uncertainty is a potential key uncertainty and should be evaluated with a realistic sensitivity analysis.

Realistic Sensitivity Assessment Option

The analyst should select reasonable options for alternate models for the particular issue and quantify alternate basic event values. Then for each reasonable alternate model, the analyst should requantify the base PRA using the alternate basic event value and the relationship between the maximum acceptable CDF (CDF^+) and reevaluate the new CDF estimate as expressed in Equation 5-4. The multiple CDF values generated using the alternate models create a range of potential base PRA results that are compared to the acceptance criteria. If any of the results would lead to not meeting the acceptance guidelines, then the issue is considered as a key uncertainty.

5.5.2.3 Case 1c: Sources Linked to the Logic Structure of the PRA

The sources of model uncertainty and related assumptions identified in Step 2 are reviewed to determine those that are relevant only to the logic structure of the PRA.

Alternative methods or assumptions that could possibly introduce new cut sets in existing sequences by changing fault tree models, new sequences by changing the structure of event trees, or even entirely new classes of accident sequences by introducing new initiating events need to be assessed by manipulating or altering the PRA to reflect these alternatives. Once again, reasonable uncertainty analyses for issues should be conducted for each logical grouping to evaluate whether or not the issues are key sources of uncertainty.

Conservative Screening Option

The effort to change the PRA logic structure can involve significant resources. However, in some cases, it may be possible to perform an approximate bounding evaluation (see Section 6) that can demonstrate that the potential impact of the alternate assumption or model will not produce a result that challenges the decision criteria. As an example, this demonstration can be achieved if the effect of the model uncertainty or related assumption is limited to the later branches of the lower frequency accident sequences and the frequency of the portion of the sequences up until the branch points is low enough.

---

[12] It is important to note that the option of setting a basic event value to logically **TRUE** rather than 1, available in many PRA software packages, is not advised. Doing so risks losing information when the PRA is solved because the use of **TRUE** eliminates entire branches of fault trees and, hence, results in the loss of cut sets from the final solution.

Realistic Sensitivity Assessment Option

The analyst should select reasonable options for alternate models for the particular issue and make the required changes to the PRA logic model. Then for each reasonable alternate model, the analyst should requantify the base PRA and reevaluate the relationship between the maximum acceptable CDF (CDF^+) and the new CDF estimate as expressed in Equation 5-4. The multiple results create a range of potential base PRA results. If any of the results would lead to not meeting the acceptance guidelines, then the issue is a key uncertainty.

5.5.2.4 Case 1d: Sources Linked to Logical Combinations

The sources of model uncertainty and related assumptions identified in Step 2 are reviewed to determine those that are relevant to combinations of basic events and logic structure. These combinations need to be identified. Guidance is provided for determining the logical combinations, and for performing conservative and realistic sensitivity analyses.

Logical Combinations

For these cases, the combination may impose a synergistic impact upon the uncertainty of the PRA results. The resulting uncertainty from their total impact may be greater than the sum of their individual impacts. For example, several issues could relate to the same dominant cut sets, or certain sequences, or a particular event and the success criteria for systems on that event tree, or to the same plant damage state. In other words, such issues overlap each other by impacting jointly on the same parts of the risk profile modeled in the PRA. Thus, to accurately assess the full potential for the impact of uncertainty, such issues also should be grouped together.

A simple example can be found in the relationship of two models, recovery of offsite power and recovery of failed diesel generators, to the overall uncertainty of the model. Both models represent the failure to restore ac power to critical plant systems through different but redundant power sources. Hence, the potential total impact of uncertainty associated with the function of supplying ac power to emergency electrical buses would involve a joint assessment of the uncertainty associated with both models. Another example is with regard to the interaction between uncertainties associated with the dc battery depletion model and those associated with the operator actions to restore power; specifically the interrelationship between operator performance and the performance of key electrical equipment under harsh conditions (e.g., smoke, loss of room cooling). How long dc batteries can remain sufficiently charged and successfully deliver dc power to critical components relies on the shedding of nonessential electrical loads, which is achieved through the actions of reactor operators and equipment operators who operate electromechanical equipment such as motor control centers and circuit breakers as well as through procedures and the availability of required tools (e.g., lighting, procedures, communication devices). Uncertainty associated with these operator actions and the potential harsh environmental impacts on both operators and equipment should be jointly assessed for a perspective on the potential total impact of uncertainty upon the dc battery depletion model.

Moreover, the choice of HRA method can impact the uncertainty of PRA results in several areas. An interface exists between the human actions necessary to restore diesel generator operation after either failing to start or run and the time to dc battery depletion. Many diesel generators depend on dc power for field flashing for successful startup. If equipment operators fail to successfully restore diesel generator operation before the dc batteries become depleted, then the diesel generators cannot be restored to operation. Hence, the potential impact of uncertainties associated with the HEPs in the model to recover failed diesel generators and uncertainties associated with the dc battery depletion model should be assessed together.

In the above examples, the uncertainty issues were linked by their relationship to a given function, namely establishing power. However, uncertainties related to different issues also can have a synergistic effect. As an example, consider the case of an uncertainty associated with the modeling of high-pressure coolant injection (HPCI) in a PRA for a boiling-water reactor. In core damage sequences of transient event trees, failure of the HPCI is either coupled with failure of other high-pressure injection systems (reactor core isolation cooling [RCIC], recovery of feedwater, control rod drive [CRD]) and failure of depressurization, or failure of other high-pressure injection systems and failure of low-pressure injection. The importance of HPCI is affected by the credit taken for additional injection systems (over and above RCIC). For example, taking credit for fire water (as an additional low-pressure system), CRD, or recovery of feedwater (as a high-pressure system) can lessen the importance of HPCI.

In the LOOP/station blackout (SBO) tree, a significant function of HPCI is to provide a delay to give time to recover the offsite power. Therefore, the modeling of recovery of offsite power in the short term (given that HPCI has failed), the frequency of LOOP, and the common cause failure (CCF) probability of the diesels and the station batteries all have an impact on the importance of HPCI.

HPCI importance is therefore affected by the following:

- Transient frequencies.

- HEP for depressurization.

- Credit for motor-driven feedwater pumps.

- Credit for alternate injection systems (e.g., fire water, service water cross-tie).

- LOOP frequency, CCF of diesels and batteries, and the factors associated with the short-term recovery of ac power given a LOOP.

In general, uncertainties associated with any of these issues could interact synergistically to impact the overall model uncertainty associated with the modeling of the HPCI.

Once the various combinations have been identified, the quantitative assessment can be performed.

Further guidance on grouping issues into logical groupings can be found in Section 4.3.2 of the EPRI report. The analyst's judgment and insight regarding the PRA should yield logical groupings specific to the PRA in question. Certain issues may readily fall into more than one logical grouping depending on the nature of the other issues.

Conservative Screening Option

When all the contributors to the logical group of sources of uncertainty only impact basic events, the approach is similar to Case 1b with regard to quantitative screening. The concept of setting all relevant basic events to 1 simultaneously and then reevaluating the PRA yields the same perspective for a logical grouping of sources of uncertainties as for a single source of uncertainty that impacts several basic events. Hence, by setting all basic events relevant to a logical grouping of sources of uncertainty to 1 to calculate a CDF^+_j where j represents the set of basic events relevant to the jth logical grouping of sources of uncertainty, Equation 5-4 can be evaluated to determine if the logical grouping is potentially a key uncertainty. If Equation 5-4 is true, then the sources of uncertainly included in the logical grouping do not present potential key sources of uncertainty as no mathematical possibility exists that the values of the basic events linked to the particular logical grouping could achieve an unacceptably high CDF. Otherwise, the sources included in the logical group of sources of uncertainty are potential key sources of uncertainty and should be evaluated with a realistic sensitivity analysis.

If the logical combinations involve both impacts on basic event values and the PRA structure, the process of performing conservative screening becomes more difficult. The same process as identified above for case 1c can be performed to address the affect of the model uncertainty on the PRA logic. The alternate PRA structure then can be used in conjunction with the process identified above to assess the affect of the alternate model on multiple basic events and the PRA structure.

Realistic Sensitivity Assessment Option

The analyst selects reasonable options for alternate models for the particular issues in the logical grouping and modifies the PRA logic and basic event values based on the selected alternate model. Then, for each reasonable alternate model, the PRA is requantified. The multiple results create a range of potential base PRA results. If any of the results would lead to not meeting the acceptance guidelines, then the logical grouping is considered as a group of potentially key sources of uncertainty.

5.5.3 Case 2: Applications Involving a Two-Metric Acceptance Guideline

Guidance is provided in this section for performing screening or assessment for each of the four classes of screening analysis for applications involving a two-metric acceptance guideline. In general, these types of applications are license amendment applications. For example, quantitative assessment of the risk impact in terms of changes to the CDF (i.e., ΔCDF) (or LERF)[13] metric is used in comparison against the RG 1.174 acceptance guidelines (Figure 5-4) or guidelines derived from those of RG 1.174.

---

[13] The discussion that follows is in terms of CDF but can also be applied to LERF.

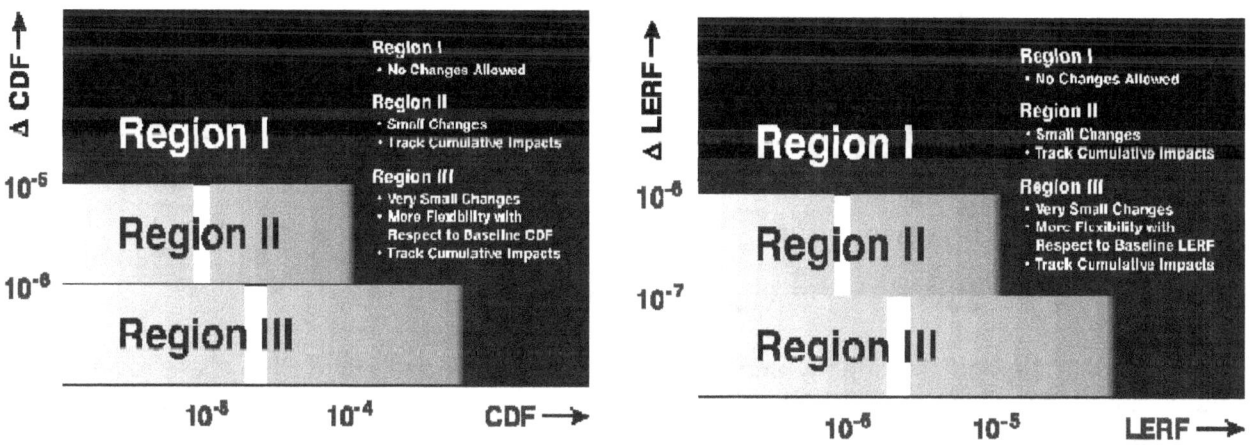

Figure 5-4   NRC RG 1.174 acceptance guidelines for CDF and LERF

Because the acceptance guidelines involve two metrics (CDF on the horizontal axis, and ΔCDF on the vertical axis) it is necessary to assess the potential impact of a model uncertainty issue not only with respect to CDF but to ΔCDF as well because acceptability result is based on the relative position within the figure. Hence, just as for applications involving only the base PRA (see Section 5.4.2), one is interested in assessing the potential impact of model uncertainty upon CDF_{base}; however, it also is an issue that needs to be assessed for its impact upon ΔCDF. Therefore, the following metrics are of interest for applications involving a change to the licensing basis:

CDF_{base} the value of the CDF mean estimate in the base PRA

CDF_{after} the value of the CDF mean estimate in the modified base PRA to account for changes proposed to the licensing basis

$CDF^+_{j,\ base}$ the CDF mean estimate in the base PRA with the basic event j set to 1

$CDF^+_{j,\ after}$ the CDF mean estimate in the modified PRA with the basic event j set to 1

Using these four quantities, the terms ΔCDF and ΔCDF$^+$ are defined as follows:

$$\Delta CDF = CDF_{after} - CDF_{base}$$ Equation 5-5

$$\Delta CDF^+_j = CDF^+_{j,after} - CDF^+_{j,base}$$ Equation 5-6

Equations 5-5 and 5-6 allow for the assessment of the potential vertical movement into unacceptable regions of RG 1.174 acceptance criteria. Such movement would identify the model uncertainty or related assumption as a potential key source of uncertainty, which would add significance to the regulatory decision at hand.

Guidance is provided for each of the four cases discussed earlier in Section 5.5, where the sources of model uncertainty or related assumptions are linked to:

- A single basic event.
- Multiple basis events.
- The logic structure of the PRA.
- Logical combinations.

5.5.3.1 Case 2a: Sources Linked to a Single Basic Event

The sources of model uncertainty and related assumptions identified in Step 2 are reviewed to determine those which are relevant only to a single basic event. For each identified source of uncertainty, a conservative screening or a realistic sensitivity analysis is performed.

Conservative Screening Option

The terms on the right-hand side of Equations 5-5 and 5-6 are readily calculable. In Equation 5-5, CDF_{base} and CDF_{after} are calculated using the base PRA and the modified PRA, respectively. In Equation 5-6, the base and modified PRAs are recalculated with the value of the relevant basic event (or the jth basic event) set to 1 in both the base and modified PRAs. Exercising the base and modified PRAs to calculate the right-hand terms of Equations 5-5 and 5-6 and then solving Equations 5-5 and 5-6 for ΔCDF and ΔCDF^+, respectively, allows for the plotting of the two ordered pairs shown in Table 5-2 against the acceptance guidelines from RG 1.174 in Figure 5-4.

Table 5-2   Ordered pairs of CDF and ΔCDF and comparison against acceptance guidelines.

Ordered pair	Perspective of comparison
$(CDF_{base}, \Delta CDF)$	Comparison of the mean CDF and mean ΔCDF against the acceptance guidelines. Provides the analyst's best judgment of the impact of the change in risk.
$(CDF^+_{j,base}, \Delta CDF^+_j)$	Comparison of the greatest possible shift in the base CDF and the greatest possible shift in the ΔCDF, as defined with the jth basic event quantified as 1, against the acceptance guidelines. Provides a perspective on the potential shift in both the ΔCDF and CDF value resulting from an alternate model or assumption.

A source of uncertainty can influence a decision by moving the ordered pair into a different region of Figure 5-4. For those sources of uncertainty associated with the modeling of the impact of the change associated with the application, this can only occur by changing ΔCDF. However, sources of uncertainty associated with the base PRA can impact both the CDF and the ΔCDF.

If the ordered pair associated with the source of model uncertainty were to lie in a region of the acceptance guideline that could affect the decision, the issue is potentially key and should be assessed with a reasonable sensitivity analysis. Examples of a change to the decision include not accepting the change if the result moves into Region I or introducing compensatory measures if the sensitivity study moves the result from Region III into Region II.

The significance of the ordered pair ($CDF^+_{j, base}$, ΔCDF^+_j) is that it gives the decisionmaker a perspective of the combined impact of an issue on both the base PRA and the modified PRA. An issue that might be perceived as not influencing the decision on the basis of where the ordered pair (CDF_{base}, ΔCDF) lies on Figure 5-4 might be perceived otherwise if the combined impact of the issue is factored into the decision.

Another method to assess the potential impact of a model uncertainty relevant to a single basic event is the method of Reinert and Apostolakis [Reinert, 2006]. Reinert and Apostolakis employ a method wherein they define the concept of a "threshold RAW" value (analogous to the use of RAW_{max} in Case 1a) for basic events with regard to both CDF and ΔCDF. Their definition of RAW with regard to CDF is directly from standard PRA practice:

$$RAW_{j,CDF-base} = \frac{CDF^+_{j,base}}{CDF_{base}} \quad \text{and} \quad RAW_{j,CDF-after} = \frac{CDF^+_{j,after}}{CDF_{j,after}} \quad \text{Equation 5-7}$$

RAW with regard to CDF is defined as:

$$RAW_{j,\Delta CDF} = \frac{\Delta CDF^+_j}{\Delta CDF} \quad \text{Equation 5-8}$$

where CDF and CDF^+_j are defined as in Equations 5-5 and 5-6. Substituting Equations 5-5 and 5-6 into Equation 5-8 yields:

$$RAW_{j,\Delta CDF} = \frac{CDF^+_{j,after} - CDF^{'+}_{j,base}}{CDF_{after} - CDF_{base}} \quad \text{Equation 5-9}$$

Solving the relationships in Equation 5-7 for $CDF^+_{j, after}$ and $CDF^+_{j, base}$ and inserting the results into Equation 5-9 yields

$$RAW_{j,\Delta CDF} = \frac{\left(RAW_{j,CDF-after}\right)x\left(CDF_{after}\right) - \left(RAW_{j,CDF-base}\right)x\left(CDF_{base}\right)}{CDF_{after} - CDF_{base}} \quad \text{Equation 5-10}$$

The right-hand terms of Equation 5-10 are readily calculable, which allows the analyst to calculate a RAW with regard to CDF. Reinert and Apostolakis use the relationships in Equations 5-8 and 5-10 to calculate threshold RAWs with regard to both CDF and CDF by selecting maximum acceptable values for CDF and CDF (which they refer to as $CDF_{threshold}$ and $\Delta CDF_{threshold}$, respectively) and then substitute these threshold values for $CDF^+_{j,base}$ and CDF^+_j in Equations 5-7 and 5-8 to yield

$$RAW_{CDF,threshold} = \frac{CDF_{threshold}}{CDF_{base}} \quad \text{Equation 5-11}$$

$$RAW_{\Delta CDF,threshold} = \frac{\Delta CDF_{threshold}}{\Delta CDF}$$

Equation 5-12

Equations 5-11 and 5-12 yield threshold values for the RAW with regard to the base PRA CDF defined in Equation 5-7 and the RAW with regard to CDF defined in Equation 5-10. The base PRA and the modified PRA are exercised, which yields $RAW_{j,\ CDF\text{-}base}$ values for all basic events in the base PRA and allows for the solving of Equation 5-10 to calculate $RAW_{j,\ \Delta CDF}$ values. The resulting values for $RAW_{j,cdf\text{-}base}$ and $RAW_{j,\Delta CDF}$ are compared to the threshold values calculated by Equations 5-11 and 5-12 to determine if any model uncertainty associated with a single basic event poses a potential key model uncertainty.

In employing the method of Reinert and Apostolakis, care should be given with regard to assessing the potential combined impact of a model uncertainty on both CDF and ΔCDF. This method does not automatically investigate the potential that an acceptable impact upon CDF and an acceptable impact upon ΔCDF could nonetheless result in an overall result that is not favorable, which is the function of the order pairs $(CDF^{+}_{j,base}, \Delta CDF)$ and $(CDF^{+}_{j,base}, \Delta CDF^{+}_{j})$ in the ordered pair approach discussed above. Reinert and Apostolakis do address this issue by selecting more than one threshold value for CDF and ΔCDF based on the horizontal and vertical transitions between regions I and II and between regions I and III.

Reinert and Apostolakis provide a case study to illustrate this method.

Realistic Sensitivity Assessment Option

The terms for the ordered pairs in Table 5-2 are evaluated for any reasonable hypothesis developed for any source of model uncertainty or related assumption linked to the jth basic event. For any such reasonable hypothesis, if any of the ordered pairs in Table 5-2 yields a result in or close to region I, then the source of model uncertainty or related assumption is a key uncertainty.

As discussed above, Reinert and Apostolakis provide an alternate method to test for whether or not a source of model uncertainty or a related assumption linked to a single basic event constitutes a potential key uncertainty. The application of that method is continued here. To perform a reasonable sensitivity analysis on model issues that can be related to specific basic events, Reinert and Apostolakis continue to employ the concept of a threshold RAW. The term "threshold" RAW importance measure coined by Reinert and Apostolakis also can be thought of as a maximum acceptable RAW importance measure in that it represents the largest possible value for the RAW importance measure for which it would be mathematically impossible for the uncertainty associated with a particular basic event to cause a PRA result that would move from one region of the figure to another. Threshold values for the RAW_{cdf} and the $RAW_{\Delta cdf}$ are calculated as[14]

$$RAW_{CDF,threshld} = \frac{CDF_{threshold}}{CDF_{base}}$$

Equation 5-13

---

[14] These are Equations 12 and 13 in Reinert and Apostolakis.

$$RAW_{\Delta CDF, threshold} = \frac{\Delta CDF_{threshold}}{\Delta CDF}$$

Equation 5-14

where CDF$_{threshold}$ is the value of CDF that corresponds to the vertical line between the applicable regions in Figure 5-5, and ΔCDF $_{threshold}$ is the value of ΔCDF that corresponds to the horizontal line between the applicable regions in Figure 5-4.

Once the RAW$_{cdf, threshold}$ and the RAW$_{\Delta cdf, threshold}$ have been calculated, each model issue can be evaluated by investigating the model for the basic events relevant to each issue. For any particular issue, the value for a particular relevant basic event j is adjusted and both the base PRA and the modified PRA are reevaluated until a result for the base PRA is found that yields one of the following:

$$RAW_{j,CDF} \approx RAW_{CDF,threshold}$$

Equation 5-15

$$RAW_{j,\Delta CDF} \approx RAW_{\Delta CDF,threshold}$$

Equation 5-16

If the value of the jth basic event that corresponds to the approximation in either Equation 5-15 or 5-16 is based on a reasonable hypothesis for the basic event's probability, then the source of uncertainty linked to the jth basic event is a key uncertainty. If this is the case, the sensitivity analysis should continue so that a "reasonable" maximum value for the jth basic event and its corresponding "high" estimates for CDF$_j$ and CDF$_j$ are calculated. These values for CDF$_j$ and CDF$_j$ most likely will be less than the values of CDF$^{+}_{j, base}$ and CDF$^{+}_{j}$, respectively, that were calculated by setting the value of the jth basic event to 1. These "reasonable" high values for CDF$_j$ and ΔCDF$_j$ will be necessary for the comparison of the risk-informed application to the RG 1.174 acceptance criteria. If the value of the jth basic event that yields the approximation in either Equation 15 or 16 is based on an unreasonable hypothesis, then the issue is not a key uncertainty because the uncertainty associated with the issue could not reasonably affect the decision.

5.5.3.2 Case 2b: Sources Linked to Multiple Basic Events

The sources of model uncertainty and related assumptions identified in Step 2 are reviewed to determine those that are relevant to multiple basic events. An example would be the choice of model to quantify human errors and recovery actions, or an assumption that affects the quantification of a particular failure mode of several redundant components (e.g., stuck-open safety relief valve). For each identified source of uncertainty, a conservative screening or a realistic sensitivity analysis is performed.

Conservative Screening Option

The RAW importance measures for several basic events cannot be used collectively to assess the combined impact of the uncertainties associated with the group of basic events. However, the concept of setting all relevant basic events to 1 simultaneously and then reevaluating the PRA yields the same perspective for a group of basic events as does the RAW importance measure for an individual basic event. Hence, all basic events relevant to a particular source of uncertainty are set to 1 to calculate the ordered pairs in Table 5-2, where j would connote the set of basic events relevant to the jth source of uncertainty. If the ordered pair associated with the source of model uncertainty were to lie in a region of the acceptance guideline that could affect the decision, the issue is potentially key and should be assessed with a reasonable sensitivity analysis.

Realistic Sensitivity Assessment Option

The analyst selects reasonable options for alternate models or assumptions for the particular issue. Then, for each reasonable alternate model, the base and modified PRAs are requantified using basic event values generated from the alternate models. The multiple results create a range of potential base PRA and modified PRA results. If any of the results yield values for the terms in the ordered pairs of Table 5-2 such that the plotting of any of the ordered pairs against the acceptance criteria of Figure 5-5 could affect the decision, then the issue is a key uncertainty.

5.5.3.3 Case 2c: Sources Linked to the Logic Structure of the PRA

The sources of model uncertainty and related assumptions identified in Step 2 are reviewed to determine those that are relevant to the logic structure of the PRA. For each identified source of uncertainty, a conservative screening or a realistic sensitivity analysis is performed.

Alternative methods or assumptions that could possibly introduce new cut sets in existing sequences by changing fault tree models, new sequences by changing the structure of event trees, or even entirely new classes of accident sequences by introducing new initiating events need to be assessed by manipulating or altering the PRA to reflect these alternatives. New estimates for $CDF^+_{j,base}$ and $CDF^+_{j,after}$ can be developed, where these terms are now defined as follows:

$CDF^+_{j,base}$ The base PRA CDF mean estimate where the base PRA has been modified to address the jth source of model uncertainty or related assumption that is linked to the logic structure of the PRA.

$CDF^+_{j,after}$ The base PRA CDF mean estimate where the PRA, as modified for the application, has been further modified to address the jth source of model uncertainty or related assumption that is linked to the logic structure of the PRA.

Conservative Screening Option

The effort to change the PRA logic structure can involve significant resources. However, in some cases, it may be possible to perform an approximate bounding evaluation (see Section 6) that can demonstrate that the potential impact of the alternate assumption or model will not produce a result that challenges the decision criteria. As an example, this demonstration can be achieved if the effect of the model uncertainty or related assumption is limited to the later branches of the lower frequency accident sequences, and the frequency of the portion of the sequences up until the branch points is low enough.

Realistic Sensitivity Assessment Option

The analyst selects reasonable options for alternate models for the particular issue. Then, for each reasonable alternate model, the base and modified PRAs are requantified. Using Equations 5 and 6, the analyst can calculate values for the terms of the ordered pairs in Table 5-2 and compare the plots of those ordered pairs to the acceptance criteria shown in Figure 5.5. If the ordered pair associated with a source of model uncertainty were to lie in a region of the acceptance guideline that could affect the decision, the issue is potentially key. Examples of a change to the decision include not accepting the change if the result moves into Region I or introducing compensatory measures if the sensitivity study moves the result from Region III into Region II.

5.5.3.4 Case 2d: Sources Linked to Logical Combinations

The sources of model uncertainty and related assumptions identified in Step 2 are reviewed to determine those that are relevant to combinations of basic events and logic structure. One should not, however, restrict oneself to a short list of generic logical groupings. The analyst's judgment and insight regarding the PRA should yield logical groupings specific to the PRA in question. Certain issues may readily fall into more than one logical grouping depending on the nature of the other issues. For these cases, the combination may impose a synergistic impact upon the uncertainty of the PRA results. Further guidance on grouping issues into logical groupings can be found in Section 4.3.2 of the EPRI report. See the discussion for Case 1d in Section 5.4.2.4 for examples.

Conservative Screening Option

When all the contributors to the logical group of sources of uncertainty only impact basic events, the approach is similar to Case 2b with regard to quantitative screening. The concept of setting all relevant basic events to 1 simultaneously and then reevaluating the PRA yields the same perspective for a logical grouping of sources of uncertainties as for a single source of uncertainty that impacts several basic events. Hence, all basic events relevant to a particular model issue or to a logical group of issues are set to 1 to calculate the ordered pairs in Table 5-2, where j represents the set of basic events relevant to the jth logical group of model issues. If none of the sensitivity cases associated with the logical group could affect the decision by moving the result from one region of the figure to another, for example, then the sources of uncertainly included in the logical grouping do not present potential key sources of uncertainty as no mathematical possibility exists that the values of the basic events linked to the particular logical grouping could achieve an unacceptably high CDF. Otherwise, the sources included in the logical group of sources of uncertainty are potential key sources of uncertainty and should be evaluated with a reasonable sensitivity analysis,

Realistic Sensitivity Assessment Option

The analyst selects reasonable options for alternate models for the particular issue. Then, for each reasonable alternate model, the base and modified PRAs are requantified. The multiple results create a range of potential base PRA and modified PRA results. If any of the results yield values for the terms in the ordered pairs of Table 5-2 such that the plotting of either of the ordered pairs against the acceptance criteria of Figure 5-5 could alter the decision, then the issues associated with the logical group need to be considered as potentially key sources of uncertainty and addressed in the decisionmaking.

6. COMPLETENESS UNCERTAINTY

This section provides guidance on addressing one aspect of completeness uncertainty; i.e., those contributors that have not been included in the scope of the probabilistic risk assessment (PRA) or the level of detail of the PRA. As discussed in Sections 2 and 3, the PRA should be of sufficient scope and level of detail to support the risk-informed decision under consideration. In particular, in accordance with the Commission's Phased Approach to PRA Quality [NRC, 2003e], the risk from each significant hazard group should be addressed using a PRA model performed in accordance with a consensus standard for that hazard group endorsed by the staff. A significant hazard group (e.g., risk contributor) is one whose consideration in the decision can make a difference to the decision and, therefore, needs to be factored into the decisionmaking. However, some contributors can be shown to be insignificant or irrelevant and, therefore, can be screened from further consideration. In addition, some contributors, while not significant, may have a minor impact on the decision. The guidance in this section focuses on the use of screening and conservative analyses to address non-significant contributors.

6.1 Overview

In providing guidance on screening and conservative analyses, it is important to first understand what is meant by these terms and how these analyses are used.

Screening analyses can involve limited but realistic analyses or conservative analyses. Conservative analyses also can be used to provide conservative risk estimates for contributors not addressed with a detailed model in the PRA. Conservative analyses can include a spectrum of conservative assessments ranging from assessments that are demonstrably conservative compared to a realistic evaluation all the way to truly bounding assessments that reflect the worst possible outcome. These analyses, when used appropriately, demonstrate that the scope and level of detail of the PRA is adequate for the application (i.e., that the uncertainty associated with incompleteness is not significant).

As indicated in Section 2, a risk-informed decisionmaking process integrates insights or results from deterministic and risk analyses along with considerations of defense in depth and safety margins. Generally, PRA results such as core damage frequency (CDF) and large early release frequency (LERF) are used as the risk inputs to decisionmaking. The results are typically compared to acceptance guidelines that specify the scope of risk contributors. For example, for Regulatory Guide (RG) 1.174 applications [NRC, 2002], the risk from all hazard groups and all plant operating states (POSs) should be addressed. Depending on the decision, the needed PRA scope and level of detail can vary. An existing PRA may not always match the needed scope or contain the level of detail necessary for a specific risk-informed decision. For example, the PRA may not include analysis of accidents during low-power and shutdown (LPSD) modes of operation, specific hazard groups such as seismic events, specific initiating events, specific accident sequences such as those resulting from an anticipated transient without scram, or some component failure modes (e.g., spurious component operation or pipe ruptures). For such a situation, an analyst will have the following four options:

(1) Upgrade the PRA to address the required scope or level of detail.

(2) Use a screening analysis to demonstrate that the missing scope items are not significant to the decision.

(3)   Use a conservative analysis to quantify the risk from the contributors not addressed with a detailed model in the PRA.

(4)   Modify the application such that the missing scope or level of detail is not affecting the decision.

The approach in the second and third options, namely the use of screening and bounding analyses, is the subject of this section and involves two major steps:

- Step 1: Determining the PRA Scope and Level of Detail Required to Support an Application. This step involves the process for assessing the application in terms of identifying the PRA scope and level of detail needed to support the risk-informed decision.

- Step 2: Performing Screening and Conservative Analyses. This step involves the process for determining whether the potential risk contributors not modeled can be screened or bounded.

6.2   Step 1: Determining the Required Scope and Level of Detail Required to Support Risk-Informed Decision

In this step, the analyst is determining whether the scope and level of detail is sufficient to support the risk-informed decision under consideration. The three following steps are involved:

- Understanding the decision and application.
- Establishing the needed PRA scope.
- Establishing the needed PRA level of detail.

6.2.1   Understanding the Risk-Informed Decision and Application

The required PRA scope and level of detail are determined by the decision under consideration. Consequently, the way in which the PRA is to be used is reviewed to determine the needed scope and level of detail. Sections 3.2.3 and 3.2.4 provide detailed discussions on PRA scope and level of detail, respectively.

This determination is generally accomplished by considering the cause-and-effect relationship between the application and its impact on the plant risk. A proposed application can impact multiple structures, systems, and components (SSCs) in various ways. Consequently, the application can require changes to one or more PRA technical elements. Examples of potential impacts include the following:

- Introduces a new initiating event.

- Requires modification of an initiating event group.

- Changes to a system success criterion.

- Requires addition of new accident sequences.

- Requires additional failure modes of SSCs.

- Alters system reliability or changes system dependencies.

- Requires modification of parameter probabilities.

- Introduces a new common cause failure mechanism.

- Eliminates, adds, or modifies a human action.

- Changes important results used in other applications, such as importance measures.

- Changes the potential for containment bypass or failure modes leading to a large early release.

- Changes the SSCs required to mitigate external hazards such as seismic events.

- Changes the reliability of systems used during LPSD modes of operation.

Once this determination is established, the PRA model is then reviewed to determine if it has the needed scope and level of detail defined by the application.

6.2.2 Establishing the PRA Scope

The scope of the PRA is defined in terms of the following:

- The metrics used to evaluate risk.

- The POSs for which the risk is to be evaluated.

- The types of hazard groups and initiating events that can potentially challenge and disrupt the normal operation of the plant and, if not prevented or mitigated, would eventually result in core damage, a release, and/or health effects.

Consequently, three decisions are made by the analyst: what risk characterization, what POSs, and what types of hazard groups and initiating events need to be addressed by the PRA.

The risk metrics relevant to the decision are defined by the acceptance guidelines associated with the decision. For example, for a licensing-basis change, RG 1.174 defines the risk metrics as CDF, LERF, ΔCDF, and ΔLERF. Therefore, if the acceptance guidelines use CDF, LERF, ΔCDF, and ΔLERF, then the PRA scope generally should address these risk parameters. However, circumstances may arise where the ΔCDF associated with an application is small enough that it also would meet the ΔLERF guidelines. In this case, explicit modeling of the LERF impacts of the application may not be necessary.

The impact of the application determines the POSs to be considered. The various operating states include at power, low power, and shutdown.[1] Not every application necessarily will

---

[1]POSs are used to subdivide the plant operating cycle into unique states such that the plant response can be assumed to be the same within the given POS for a given initiating event. Operational characteristics (such as reactor power level; in-vessel

89

impact every operating state. In deciding this aspect of the required scope, the SSCs affected by the application are identified. This step includes a determination of the cause-and-effect relationships between the proposed plant application and its impact on the SSCs. It is then determined if the affected SSCs are required to prevent or mitigate accidents in the different POSs and if the impact of the proposed plant change would impact the prevention and mitigation capability of the SSCs in those POSs. A plant change could affect the potential for an accident initiator in one or more POSs or reduce the reliability of a component or system that is unique to a POS or required in multiple POSs. Once the cause-and effect-relationship on POSs is identified, the PRA model is reviewed to determine if it has the scope needed to reflect the effect of the application on plant risk.

The impact of the application also determines the types of initiating events to be considered. An initiating event is the result of a challenge to the continued operation of the plant. The challenges to the plant are classified into hazard groups that are defined as a group of similar causes of initiating events assessed in a PRA using a common approach, methods, and likelihood data for characterizing the effect on the plant. Typical hazard groups for a nuclear power plant PRA include: internal events, seismic events, internal fires, internal floods, and high winds. In deciding this aspect of the required scope, the process is similar to that described above for POSs. The SSCs affected by the application are determined, and the resulting cause-and-effect relationships are identified. It is then determined if the affected SSCs are required to prevent or mitigate both internal and external events and if the proposed plant change would affect that capability. The impact of the proposed plant change on the SSCs could introduce new accident initiating events, affect the frequency of initiators, or affect the reliability of mitigating systems required to respond to multiple initiators. Once the cause-and-effect relationship on the accident-initiating events is identified, the PRA model is reviewed to determine if it has the scope needed to reflect the effect of the application on plant risk.

For example, consider an application that involves a licensing-basis change where a seismically qualified component is being replaced with a non-qualified component. If the new component's reliability for responding to non-seismic events is not changed, the non-seismic part of the PRA is not impacted and only the seismic risk need be considered; that is, the other contributors to the risk (e.g., fire) are not needed for this application. If, on the other hand, the reliability of the new component is different for responding to non-seismic events, the non-seismic part of the PRA may be impacted and, therefore, needs to be included in the scope.

As another example, if an application does not affect the decay heat systems at a plant, an evaluation of loss-of-heat removal events during low-power shutdown would not be required. However, the assessment of other events, such as drain-down events, may still be required.

If a hazard group is determined significant to a risk-informed application, the Commission has directed that the PRA used to support that application be performed against an available, staff-endorsed PRA standard for that specific hazard group [NRC, 2003e]. However, for a given application, the PRA does not need to be upgraded if it can be demonstrated that the missing scope or missing level of detail does not impact the risk insights supporting the risk-informed application. Section 6.3 discusses the methods for achieving this demonstration.

---

temperature, pressure, and coolant level; equipment operability; and changes in decay heat load or plant conditions that allow new success criteria) are examined to identify those relevant to defining POSs. These characteristics are used to define the states, and the fraction of time spent in each state is estimated using plant-specific information.

6.2.3 Establishing the PRA Level of Detail

A number of decisions made by the analyst determine the level of detail included in a PRA. These decisions include, for example, the structure of the event trees, the mitigating systems that should be included as providing potential success for critical safety functions, the structure of the fault trees, and the screening criteria used to determine which failure modes for which SSCs are to be included.

The technical requirements of the American Society of Mechanical Engineers (ASME) and American Nuclear Society (ANS) PRA Standard [ASME/ANS, 2009] determine the minimum level of detail for a base PRA, independent of an application. The use of screening analyses (either qualitative or quantitative) is an accepted technique in PRA for determining the level of detail included in the analysis. Section 6.3.3 includes a number of examples of screening analyses. However, it is recognized that the detail specified by the technical requirements in the AMSE/ANS PRA standard may not be needed for a given application, although the minimum level of detail may not be sufficient in other cases. The level of detail needed is that detail required to capture the effect of the application (i.e., the PRA model needs to be of sufficient detail to ensure the impact of the application can be assessed). Again, the impact of the application is achieved by reviewing the cause-and-effect relationship between the application and its impact on the PRA model.

6.3 Step 2: Performing Screening and Conservative Analyses

At this step in the process, the scope and level of detail has been demonstrated to be insufficient, and the analyst is determining whether the potential risk contributors not modeled can be screened or bounded. Guidance is first provided on screening and conservative analyses, and then examples are provided to illustrate how to implement the guidance.

6.3.1 Screening Analysis

A screening analysis is performed to demonstrate that a particular item (e.g., a hazard group, an initiating event, a component failure mode, etc.) can be eliminated from further consideration in a PRA being used to support a risk-informed application. This screening can be accomplished by showing that either the item has no bearing on the application (qualitative screening) or that the contribution of the item to the change in risk associated with the application is negligible (quantitative screening).

Qualitative Screening

Many hazards, component failure modes, and other events can be qualitatively eliminated from a base PRA. In the case of a hazard group, for example, this qualitative screening can be done by demonstrating that the hazard will not result in a challenge to the plant such that either an initiating event or damage to the mitigating systems required to respond to an initiating event can occur. The PRA standard provides criteria for performing qualitative screening. Section 6.3.3 lists examples of these screening criteria.

For risk-informed applications, the cause-and-effect method of determining the scope and level of detail of the PRA outlined above is a qualitative screening process. It identifies what SSCs, POSs, hazards, and the level of detail that may be needed in the PRA for assessing the risk impact of the application. If a scope or level of detail item cannot be screened qualitatively, it may be possible to screen it quantitatively.

Quantitative Screening

A quantitative screening assessment can consist of either (1) a realistic but limited quantitative analysis or (2) a conservative or bounding quantitative analysis. In either case, the process of quantitative screening should minimize the likelihood of omitting any significant risk contributors. This is accomplished by utilizing suitably low quantitative screening criteria. The quantitative screening process needs to address the effect of the scope or level of detail item being screened on both CDF and LERF.

Quantitative Screening -- Realistic

Quantitative screening analysis can be achieved using a limited, but realistic (i.e., best estimate) risk assessment that provides resulting risk values that are below the acceptance criteria associated with the decision. A process of successive screening using the PRA technical elements as a guide can be used in this effort. A progressive screening process may begin with screening out an initiating event with a sufficiently low frequency and proceed to screening specific sequences if the initiating event can not be screened. The progressive screening approach may require a detailed best estimate analysis of the item that may or may not result in screening.

Quantitative Screening -- Conservative

Alternatively, the quantitative screening can be performed using models and data values that are demonstrably conservative compared to best estimate versions. Different levels of conservatism can be employed with the highest level resulting in a bounding risk estimate. For example, use of conservative human error probabilities may be used in a risk evaluation, but a bounding risk estimate would require that the human error probabilities be set to 1. The following section further discusses the difference between conservative and bounding analyses.

As with the use of best estimate models, conservative models can be applied in a progressive screening approach. If the item is not screened using a conservative analysis, the results of the conservative analysis can be used in the application as a conservative or bounding risk estimate if it is not significant to the decision. Any approximations or simplifications used in the screening process needs to result in conservative or bounding risk estimates. It also is important to examine the interaction of different conservative models and data to ensure that the overall results are conservative. Finally, it is possible that a specific item could be screened using a combination of conservative and best estimate models and data.

6.3.2 Conservative Analysis

A conservative analysis may be used for one of the following two purposes with regard to scope or level of detail missing from a PRA used in an application:

- To screen the risk contribution of a PRA item from further consideration in the risk assessment of the application (discussed in the previous section).

- To bound the risk contribution for consideration in the risk-informed application decision when the scope or level of detail item can not be screened.

In the first situation, a conservative or bounding analysis is used as part of the screening analysis to demonstrate that the risk from the scope (or level of detail) is not a contributor. In the second situation, a conservative or bounding analysis is used in lieu of a detailed PRA model or a detailed element of a PRA model to estimate the risk contribution from the item.

To screen out a contributor, the conservative analysis needs to demonstrate that the impact on the PRA model is essentially negligible. When used as a surrogate for a detailed model, the impact on risk needs to be shown to be insignificant with respect to the results being used to support the decision.

In the context of a specific scope item (i.e., hazard group or POS), a bounding analysis, in general, is one that captures the worst credible outcome of a set of outcomes that result from the item being assessed. The scope item being assessed is the risk contributor from the PRA (e.g., the risk from anticipated transient without scram scenarios, the risk from internal fire events). The worst outcome is the one that is the most challenging to the plant and has the greatest impact on the defined risk metric(s). Moreover, a bounding analysis should be bounding both in terms of the potential outcome and the likelihood of that outcome. Consequently, a bounding analysis in general considers both frequency of the event and outcome of the event.

This definition can, in principle, be applied to a level of detail item. Further, this definition is consistent with, but more inclusive than, that provided in the ASME/ANS PRA Standard, which defines a bounding analysis as "analysis that uses assumptions such that the assessed outcome will meet or exceed the maximum severity of all credible outcomes."

How the above definition is applied is dependent on whether a bounding analysis is to bound the risk or to screen the item as a potential contributor to risk. If a bounding analysis is being used to bound the risk (i.e., determine the magnitude of the risk of an event), then both its frequency and outcome are considered. However, if a bounding analysis is being used to screen the event (i.e., demonstrate that the risk from the event does not contribute to the defined risk metric[s]), then it can be screened on frequency, outcome, or both, depending on the specific event.

In contrast, a conservative analysis provides a result that may not be the worst of the set of outcomes but clearly represents a result that would be significantly greater than what would be obtained from a best-estimate evaluation. An analyst can employ different levels of conservatism to address specific scope items with the highest equal to that provided by a bounding analysis. The degree of conservatism is determined by the selection of the models and data as well as the degree of modeling used to analyze the scope item. For example, all seismic events could conservatively be assumed to result in loss of offsite power (LOSP) transients or LOSP events combined with a small loss-of-coolant accident (LOCA). For screening purposes, the level of conservatism utilized is generally the minimum required to generate a frequency, consequence, or risk estimate that is below established criteria. Because a less-than-bounding but conservative analysis may not result in screening, a more detailed analysis may need to be performed to either screen the scope item or to provide a more realistic estimate for use in the risk-informed application.

Given the above discussion on what constitutes a conservative and bounding analysis, a determination needs to be made concerning the acceptability of either the conservative or bounding analysis. The criteria for acceptability of a conservative analysis are related to the type of event being bound or screened. The analyst's conservative assumptions will be specific to the event under consideration. Nonetheless, each conservative analysis needs to address the following to be acceptable:

- Completeness of potential impacts and their effects.
- Frequency.

Completeness of Potential Impacts and Their Effects

The spectrum of potential impacts of the missing scope or level of detail item and the effects on the evaluation of risk has been addressed such that impacts or effects that could lead to a more severe credible outcome have not been overlooked.

For example, suppose that the PRA did not initially address LOCAs. If a conservative analysis were to be used to address LOCAs, the full spectrum of break sizes would need to be considered. If the spectrum only included break sizes from 8 inches to 24 inches, the analysis would not be bounding because the break sizes not considered (e.g., greater than 24 inches) could have a more severe outcome.

The different effects (e.g., accident progression) resulting from the events should be addressed such that a more severe credible outcome is not overlooked. That is, the different accident progressions have been identified and understood to the extent that a different, more severe credible outcome is unlikely.

Frequency

The frequency used in the conservative analysis should be greater than, or equal to, the maximum credible collective frequency of the spectrum of impacts analyzed for the missing item. For example, if large LOCAs were missing from the PRA scope, a bounding assessment of the missing range of LOCAs would have to utilize the total frequency for the spectrum of LOCAs that are not in the PRA.

6.3.3   Examples Selecting and Using Screening and Conservative Approaches

Although an acceptable conservative analysis has general characteristics as provided above, the acceptability of a conservative analysis also is dependent on the type of event being analyzed (i.e., the analyst's conservative or bounding assumptions will be specific to the event under consideration). This dependency is illustrated in the examples of various applications of conservative analyses discussed below. The following subsection first presents examples of conservative analyses used to screen risk contributions from further consideration and then illustrates some examples of analyses used to bound the risk contribution to be included in the risk estimate.

6.3.3.1   Examples of Screening of Risk Contributors

The general process for the screening of PRA scope items is a progressive process that can involve different levels of qualitative and/or quantitative screening. In general, qualitative screening is performed prior to any quantitative screening analysis. Qualitative screening generally involves an argument that the scope item cannot impact plant risk or is not important to the change in risk (i.e., CDF or LERF) associated with the proposed plant modification. An example of the former is that the potential for specific hazard groups can be eliminated for many plant sites. An example of the latter is that changing an at-power technical specification does not impact LPSD risk. Quantitative risk screening involves a conservative estimate of the risk or change in risk from the proposed plant modification related to a scope item. Examples of this include analyses that show an initiating event has a very low frequency or analyses that indicate a plant change does not significantly change the unavailability of a system.

Both qualitative and quantitative approaches can utilize conservative deterministic analyses to support the screening process. An example of a deterministic analysis in qualitative screening is the use of a conservative thermal-hydraulic evaluation that shows the conditions for a phenomenon such as pressurized thermal shock cannot occur for a given scenario, eliminating the need to include vessel rupture due to this failure mechanism in the risk evaluation. Another example is the performance of a structural analysis that indicates a building cannot be damaged by an external hazard such as an explosion.

An example of a deterministic analysis supporting a quantitative screening process is provided in the quantitative fire compartment screening process documented in NUREG/CR-6850, "EPRI/NRC-RES Fire PRA Methodology for Nuclear Power Facilities," [EPRI, 2005]. This shows where conservative fire modeling analyses are used to identify fire sources that can potentially cause damage to important equipment. This analysis allows for eliminating fire sources that cannot cause damage, thus reducing the fire frequency for compartments that is used in the quantitative screening process.

Qualitative and quantitative arguments for screening scope items are based on the application of appropriate screening criteria. Recommended screening criteria for eliminating items from the scope of a base PRA are provided in the ASME/ANS PRA Standard. These screening criteria are applicable for base PRAs, subject to their endorsement or clarification as provided in Regulatory Guide 1.200[15] [NRC, 2007a]. In addition, screening criteria in NRC guidance documents can be utilized. Examples of these screening criteria are specified in the joint Electric Power Research Institute /NRC fire PRA methodology.

The screening process performed for a base PRA needs to be reviewed for a risk-informed application to verify that the screening is still appropriate. This can be accomplished by confirming that the proposed plant modification or operational change does not change the reason for the screening in the base PRA. The cause-and-effect relationship that is used to establish the impact of the proposed change on SSCs and the required scope of the PRA will provide the information necessary to accomplish this review. The cause-and-effect determination also can be utilized for qualitative screening of scope items. In this case, if no identified effect can be found on specific scope items, then the risk parameters used to evaluate the plant change will remain the same. However, if an identified effect is found on specific scope items, then either a screening or a detailed quantitative analysis (best estimate or conservative) is performed to determine the impact on the required risk parameter.

Qualitative Screening

Qualitative screening is utilized in many facets of establishing a base PRA. It involves the application of approved screening criteria such as those specified in PRA standards and guidance documents to eliminate potential risk contributors from the PRA. Because the performance of a base PRA is a potential application (e.g., for identification of plant

---

[15] Revision 2 to RG 1.200, which endorses ASME/ANS PRA Standard RA-Sa-2009 [ASME/ANS, 2009], is being finalized at the time of publication of this NUREG, and should be available in April 2009.

vulnerabilities as required by Regulatory Guide 1.206 [NRC, 2007b]), these screening criteria are applicable to this application. Furthermore, as previously indicated, the screening process used in the base PRA needs to be reviewed for applicability following a proposed plant change. Some examples of these criteria are provided below to illustrate the general nature of the qualitative screening criteria that can be utilized.

The primary application of qualitative screening criteria is in the analysis of spatially related initiators such as internal fire and flooding and external hazards. For example, in internal fire and flood analyses, compartments can be eliminated from analysis if they contain no equipment whose failure can cause a plant trip (manual or automatic) or require the need for an immediate plant shutdown, and no SSCs required to mitigate the event are located in the compartment. Variations of these criteria exist and, in some cases, require the performance of a deterministic analysis. For example, the requirements for internal flood areas provided in the ASME/ANS PRA Standard also allow screening of flood areas if no plant trip or shutdown will occur AND if the flood area has flood mitigation equipment such as drains capable of preventing flood levels from causing damage to equipment or structures. A deterministic analysis would be necessary in this case to show that the drains are sufficient in size to prevent the water level from all flood sources from rising to a level that would result in equipment or structure damage.

Qualitative screening also can be accomplished by considering whether a hazard source exists in an area and if a potential exists for propagation. For example, a flood area also could be screened from consideration if no identified flood sources are in the area and if floods from other areas cannot propagate to the area.

External hazard risk analysis typically begins with a screening process to eliminate hazards that are not possible at a plant or that can be combined with other events. Although the list of possible external hazards is large, most are typically screened from analysis in PRAs. External hazards included in most PRAs include both natural and manmade events. Natural external hazards most often included in a PRA are earthquakes and high winds and tornados. Manmade external hazards are evaluated less frequently, but the most typical hazards analyzed include airplane crashes, explosions at nearby facilities, and impacts from nearby transportation activities.

The ASME/ANS PRA Standard has endorsed the following set of five external hazard screening criteria:

(1)   The hazard would result in equal or lesser damage than the events for which the plant has been designed. This requires an evaluation of plant design bases to estimate the resistance of plant structures and systems to a particular external hazard.

(2)   The hazard has a significantly lower mean frequency of occurrence than another event (taking into account the uncertainties in the estimates of both frequencies), and the hazard could not result in worse consequences than the other event.

(3)   The hazard cannot occur close enough to the plant to affect it. Application of this criterion needs to take into account the range of magnitudes of the hazard for the recurrence frequencies of interest.

(4)   The hazard is included in the definition of another event.

(5)   The hazard is slow in developing, and it can be demonstrated that sufficient time exists to eliminate the source of the threat or to provide an adequate response.

Most external hazards are screened from further consideration based on Criteria 1 and 4. In general, the following external hazards need to be considered further in a risk assessment:

- Aircraft impacts.
- External flooding.
- Extreme winds and tornados (including generated missiles).
- External fires.
- Accidents from nearby facilities.
- Pipeline accidents (e.g., natural gas).
- Release of chemicals stored at the site.
- Seismic events.
- Transportation accidents.
- Turbine-generated missiles.

The ASME/ANS PRA Standard also indicates that a secondary external hazard screening step can be performed. This step allows a hazard to be screened from further analysis if the current as-built and as-operated plant conforms to the design-basis requirements for an external hazard in the 1975 Standard Review Plan (SRP) [NRC, 1975]. NRC accepted this criterion for use in the individual plant examination of external events (IPEEE) based on the judgment that the contribution to core damage frequency from an external event (i.e., hazard) that meets the criteria in the SRP would be less than10^{-6}/yr. It is noted that the SRP requires analysis of certain design-basis events that have frequencies between 10^{-7}/yr and 10^{-6}/yr. However, NUREG-1407 [NRC, 1991] indicated this criterion should not be used for screening seismic events from the IPEEE process, and the ASME/ANS PRA Standard also does not include this criterion for screening seismic events from a base PRA.

Even for non-seismic events, the analyst should use caution in applying the above criterion. One of the pitfalls in using the SRP criterion is that emphasis is placed on comparisons of event lists with the design bases of the safety-related systems and structures. However, PRAs have shown that nonsafety-related systems provide important risk contributions, and their capacities are generally not evaluated. More importantly, the possibility also exists that the magnitude of an external event may exceed the plant design basis. In fact, if the exceedence frequency for the external hazard is relatively flat around the design-basis magnitude, then the contribution from larger magnitude events could be important. In addition, a significant risk contribution from lower magnitude events is possible if the susceptibility of the plant to damage (fragility) is relatively insensitive to the magnitude of the event. To provide a convincing case that an external hazard can be excluded based on the SRP screening criterion, it may be necessary to perform some conservative estimates of the risk for both lower and higher magnitude events.

Qualitative screening criteria also are used in the evaluation of internal events, but to a lesser extent. For example, a random initiator such as a loss of heating, ventilation, and air conditioning can be screened if the event does not require the plant to go to shutdown conditions until some time has elapsed. During this time, the initiating event conditions can be detected and corrected before normal plant operation is terminated (either automatically or administratively in response to a limiting condition of operation). Deterministic analyses would be necessary to utilize this criterion.

Qualitative screening also is performed in the evaluation of pre-accident human errors to restore equipment following test and maintenance. In this case, factors such as automatic realignment of equipment on demand, the performance of post maintenance tests that would reveal misalignment, position indicators in the control room, and frequent equipment status checks all can be used as justification for screening out pre-accident human errors. However, an implied probabilistic (quantitative) argument is associated with such screening.

Quantitative Screening

Quantitative screening is utilized in most of the technical elements that comprise a base PRA. It involves the application of approved screening criteria such as those specified in the ASME/ANS PRA Standard and guidance documents to eliminate potential risk contributors from the base PRA. The screening criteria are developed and need to be correctly utilized to ensure that the screening process does not eliminate elements of a PRA model that can provide a significant contribution to the risk estimate. These screening criteria also are applicable for use in risk-informed applications and can be used for screening missing items.

Quantitative screening criteria can be either purely quantitative or incorporate both quantitative and qualitative components. These criteria can include consideration of the individual contribution of a screened item and the cumulative contribution of all screened items. Special consideration is generally given to those items that can result in containment bypass. Some of the criteria also use comparative information as a basis for screening (e.g., an item whose contribution is significantly less than the contribution from another item that results in the same impacts to the plant can be screened). Some examples of these criteria are provided below to illustrate the general nature of the quantitative screening criteria that can be utilized. Note that some of the following criteria may not be effective for new reactors that have lower CDFs.

- Initiating events can be screened if their frequency is less than 10^{-7}/yr as long as the event does not include a high consequence event such as an interfacing system LOCA, containment bypass, or reactor vessel rupture. Alternatively, an initiating event can be screened if its frequency is less than 10^{-6}/yr and core damage could not occur unless two trains of mitigating systems are failed independent of the initiator [ASME/ANS, 2009].

- A component may be excluded from a system model if the total failure probability of all the component failure modes resulting in the same effect on system operation is at least two orders of magnitude lower than the highest failure probability of other components in the same system resulting in the same effect on system operation. A component failure mode can be excluded from the system model if its contribution to the total failure probability is less than 1 percent of the total failure probability for the component, when the effect on system operation of the excluded failure mode does not differ from the effects of the included failure modes. However, if a component is shared among different systems (e.g., a common suction pipe feeding two separate systems), then these screening criteria do not apply [ASME/ANS, 2009].

- An internal flood-initiating event can be screened if it affects only components in a single system and if it can be shown that the product of the flood frequency and the probability of SSC failures given the flood is two orders of magnitude lower than the product of the non-flooding frequency for the corresponding initiating events in the PRA and the random (non-flood induced) failure probability of the same SSCs that are assumed failed by the flood [ASME/ANS, 2009].

- A flood area can be screened if the product of the sum of the frequencies of the flood scenarios for the area and the bounding conditional core damage probability is less than 10^{-9}/yr [ASME/ANS, 2009].

- A fire compartment can be screened if the CDF is less than 10^{-7}/yr and LERF is less than 10^{-8}/yr. The cumulative risk estimate (either realistic or conservatively determined) for the screened fire compartments should be less than 10 percent of the total internal events risk [EPRI 2005b].

The ASME/ANS PRA Standard also provides guidance for probabilistic analyses of seismic, high-wind and external flooding events. A set of screening criteria is provided for eliminating external event initiators, component failures, and accident sequences. For example, an initiating event can be screened out if any of the following apply:

- The event meets the criteria in the NRC's 1975 SRP (or a later revision).

- A demonstrably conservative analysis can show that the mean value of the frequency of the design-basis hazard used in the plant design is less than approximately 10^{-5}/yr and that the conditional core damage probability is less than 0.1, given the occurrence of the design-basis-hazard event.

- A demonstrably conservative analysis can show that the CDF is less than 10^{-6}/yr.

As noted previously, because a proposed plant change can change the basis for screening, the screening process used in the base PRA need to be reviewed for applicability following a proposed plant change. In addition, the type of screening criteria identified above can be used in conjunction with either a realistic but limited quantitative analysis or a conservative or bounding risk evaluation to eliminate items missing from the scope of a PRA necessary to support a risk-informed application.

One quantitative method consists of a progressive screening approach that is performed using the PRA technical elements as a guide. The screening process begins with screening out initiating events with sufficiently low frequencies and proceeds to screening specific sequences if necessary.

The draft ANS LPSD PRA Standard[16] does not provide any screening criteria for POSs and recommends that POSs be grouped together rather than screened. Grouping is recommended as opposed to screening because even if a plant only spends a short time in a POS, the conditional core damage probability may be high enough such that an important core damage contributor would be missed. However, a quantitative screening approach that can be used to screen certain specified shutdown POSs from requiring further quantitative risk assessment (in the context of a specified application) is based on demonstrating that the risk for these POSs is lower than some predetermined limiting value. The POSs that may be screened by this approach involve only the POSs in which a plant is shut down; this approach does not address POSs involving low-power operation or power transition (i.e., only POSs that span cold shutdown or refueling as defined by a plant's technical specifications). This approach starts by performing a qualitative comparison of the plant systems and configurations being examined with reference plant systems and configurations. If sufficient similarity exists, then conservative

---

[16] "American National Standard Low Power and Shutdown PRA Methodology," ANSI/ANS-58.22-200x, American Nuclear Society, June 2008.

or bounding numerical risk metric results (e.g., CDF) can be determined. If sufficient similarity does not exist, the method cannot be used. The basis for the numerical results is expected to ensure that the actual numerical value for the plant being examined is equal to (or less than) the conservative results. If the conservative result is above a specified screening value, the POS cannot be screened.

Quantitative screening can be used to either obtain an approximation of the appropriate risk metric (e.g., CDF or LERF) for a POS or to eliminate initiating events within a POS. An initiating event may be eliminated by the following two ways:

(1)   Elimination based on frequency times the fraction of time in a POS.
(2)   Elimination based on the amount of time before core damage begins.

An initiating event may be eliminated from further consideration if the product of its frequency and the fraction of time a POS occurs (per year) is less than a specified screening value. To be considered for elimination, the initiating event can not lead directly to a bypass of the containment (e.g., steam generator tube rupture).

An initiating event also may be eliminated if the time to core damage is greater than a specified time period (e.g., 24 hours) and the initiating event does not necessarily preclude or hamper potential recovery actions (e.g., a large seismic event might be expected to hamper potential offsite recovery actions, so it might not be screened out).

6.3.3.2 Examples of Conservative Risk Contributors

A conservative risk method can involve a simplified risk assessment where conservative failure probabilities for the available SSCs are estimated and combined with a conservative estimate of the initiating event frequency. The conservative estimation of the SSC failure probability can be based on bounding the known information (e.g., based on the failure probability of other similar SSCs or generic information); based on a conservative detailed analysis (e.g., by generating a fault tree for a system); or based on a conservative deterministic analysis (e.g., by determining the ultimate strength of a structure). It is critical in this process that the simplified conservative analysis reflects the as-built and as-operated plant and includes all dependencies. However, care need to be taken to ensure that the conservative results do not mask important risk contributors.

The evaluation of an external event can be bounded using a hazard analysis and some conservative or bounding assessment of the plant damage and consequences. The effects of the external event can be evaluated conservatively using upper bound or demonstrably conservative assumptions. For example, the maximum magnitude of a gas explosion could be used to determine the overpressure that would be applied to structures, the potential for missile generation, and the magnitude of a fire. The effects of the event on accident initiators and mitigating systems also can be bounded. For example, all systems in a building could be assumed damaged by an aircraft impact that is conservatively assessed to result in building damage.

Specific guidance for performing conservative assessments of external events is provided in NUREG/CR-4832 [SNL 1992a] and NUREG/CR-4839 [SNL, 1992b]. This guidance was utilized in the risk methodology integration and evaluation program study and is valid for use in risk-informed applications. It addresses aircraft crashes, high winds and tornados, transportation accidents, turbine-generated missiles, accidents at nearby facilities, and external flooding. The guidance can be used to establish the baseline risk from these external events and to determine the risk from a proposed risk-informed change to the design or operation of the plant.

In fire PRAs, a conservative assessment may be used in lieu of a detailed assessment for a physical analysis unit when it is demonstrated that the risk associated with that conservative assessment does not contribute significantly to the desired fire PRA result. This approach is codified in the ASME/ANS PRA Standard in the technical element QNS, Quantitative Screening. This term is perhaps a misnomer because the results are not screened out but retained in the PRA model as knowingly conservative estimates.

PRAs typically combine similar initiating events into groups for evaluation. For example, many transient with less restrictive system success criteria can be combined into a loss of feedwater transient group where the system success criteria may be more limiting. However, the following other factors need to be considered in combining similar initiating events besides the system success criteria to ensure that the resulting risk estimate is conservative:

- The impact of the initiating event on the SSCs needs to be included in the analysis.

- The success criteria for systems need to be more restrictive during all phases of an accident.

- Mitigating system requirements (including support systems) and dependencies for the unanalyzed events should be bounded by the event analyzed.

- The timing of the accident sequences and the impact on the operator actions needs to be bounded.

- Phenomenological conditions created by the missing accident needs to be included in the analysis.

- The potential for a large, early release is bounded by the analysis.

It may be possible to estimate the risk associated with certain LPSD POSs with a conservative analysis. These POSs include those for which the technical specifications are unchanged from power operation. If the technical specifications for a POS are the same as those at power, the CDF or LERF may be approximated by simply multiplying the fraction of time (per year) spent in the POS by the appropriate risk metric determined by the at-power PRA. This approach can be used as long as no new initiating events exist for the POS in question nor any change to the frequency of initiating events as defined in the at-power PRA. If changes to initiating event frequencies are possible, then the at-power PRA model needs to be either requantified (if the initiating frequency decreases) or resolved (if the initiating frequency increases).

7. TREATING PRA UNCERTAINTY IN RISK-INFORMED DECISIONMAKING

This section provides guidance on how to address probabilistic risk assessment (PRA) uncertainty in risk-informed decisionmaking. A major part of this guidance involves assessing the confidence in the conclusions drawn from the analysis by addressing the uncertainty in the results of the risk assessment and their contribution to the decision. This guidance is provided in the context of the integrated risk-informed decisionmaking process. In addition, this section specifies how the guidance in the previous sections is factored into this process and focuses on some special issues that need to be addressed. Consequently, the guidance includes:

- Process overview
- Process for evaluating the uncertainties in decisionmaking
- Special issues

7.1 Process Overview

As discussed in Section 2.1, an evaluation of the risk implications is one of several inputs to making a decision in a risk-informed environment. Figure 7-1, which is based on Regulatory Guide (RG) 1.174 [NRC, 2002], identifies the principles of risk-informed decisionmaking.

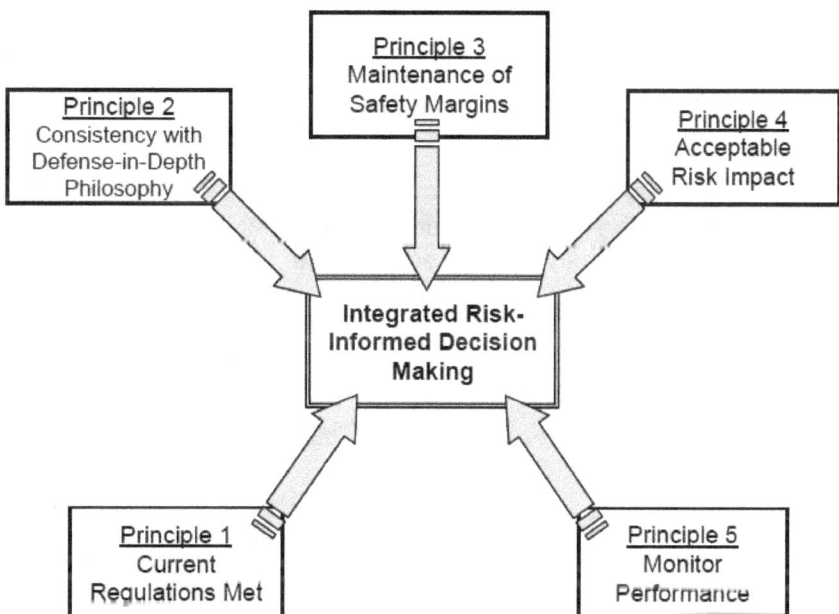

Figure 7-1 Principles of risk-informed decisionmaking

These principles, as discussed in Section 2, form the foundation of the risk-informed decisionmaking process. As discussed below, Principle 4 is of primary interest in developing the guidance for addressing PRA uncertainty in risk-informed decisionmaking.

The principles themselves are not the risk-informed decisionmaking process. Figure 7-2 provides a representation of elements of the risk-informed decisionmaking process. The dotted line delineates those parts of the process addressed in this document.

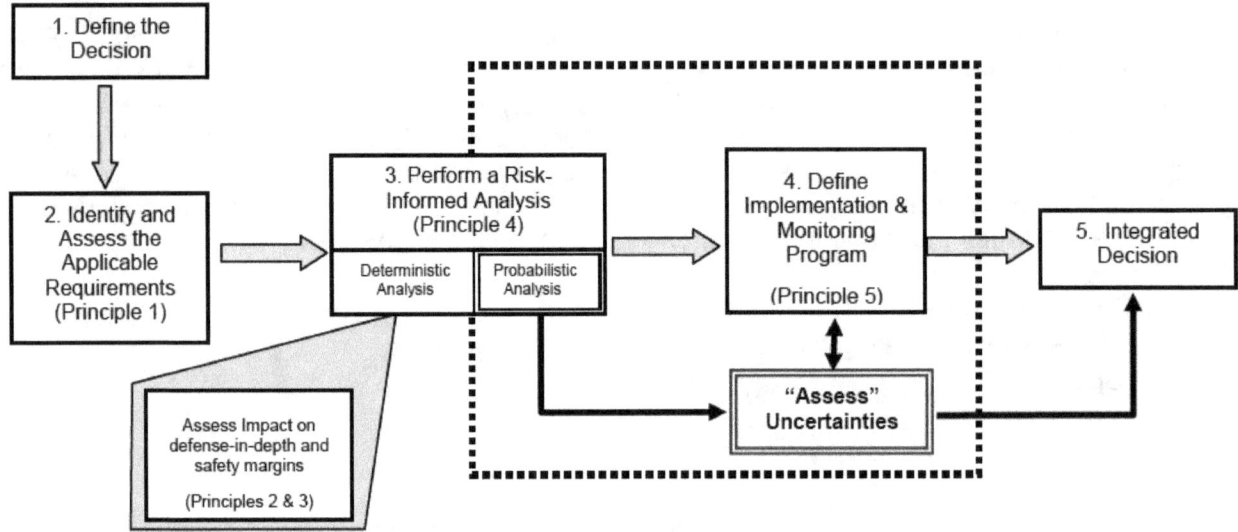

Figure 7-2 Elements of the integrated risk-informed decisionmaking process

The probabilistic analysis identified in Element 3 of the risk-informed decisionmaking process is the risk assessment that was discussed at a high level in Section 3 and is discussed in detail in this section. The PRA analysis is most relevant to addressing Principle 4 but can have a role in defining the implementation and monitoring program as discussed later. This section addresses how the PRA uncertainties are taken into account in addressing Principle 4 and how they may impact Element 4 of the integrated decisionmaking process. The guidance here does not instruct the analyst what decision to make but focuses on what information needs to be provided to the decisionmaker. Consequently, when providing input to the decisionmakers, the risk analyst is responsible for ensuring that the conclusions of the risk assessment are documented and communicated clearly. An important part of the documentation of the risk assessment is a discussion of the robustness of, or the confidence in, the conclusions drawn from that analysis.

Figure 7-3 identifies the major steps of this probabilistic analysis, and these major steps are discussed briefly below.

Initiation of the process starts with a description of the application. A significant number of potential risk-informed applications exist, and Table 3-1 identifies a number of the more common applications. For those applications identified in Table 3-1, guidance documents exist in the form of regulatory guides or industry guidance documents that specify the risk analysis methods and the acceptance guidelines to be used. This aspect is not discussed further in this document. The three steps, as shown above and described below, are the focus of the developed guidance.

104

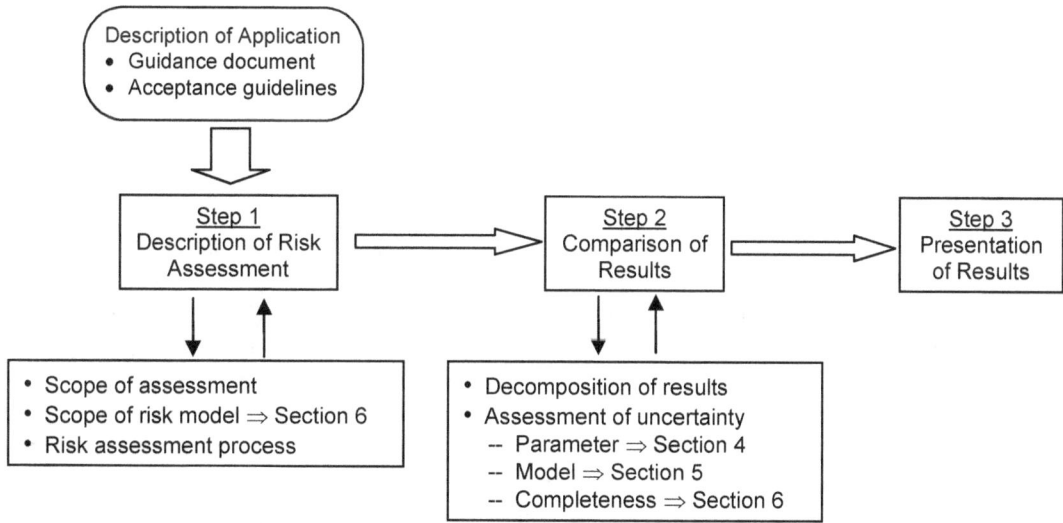

Figure 7-3 Process for evaluating the results of the risk assessment

Step 1: Description of the Risk Assessment

The required risk assessment is a function of the decision and is generally specified in the guidance documents associated with the application as discussed in Section 3. The risk assessment to be performed is specified in terms of its scope, level of detail, and how the PRA model is to be used to generate the results for the quantified risk contributors needed for comparison with the acceptance guidelines. Section 7.2 discusses this guidance in detail.

Step 2: Comparison with Acceptance Guidelines

In the context of risk-informed decisionmaking and to address Principle 4, the results of a risk assessment are compared with acceptance guidelines associated with the decision under consideration. When comparing the results of the risk assessment to acceptance guidelines, it is necessary to compare the numerical results and to understand the contributors to those results. In particular, when the results challenge the acceptance guidelines, it is important to identify if any unnecessary conservatisms are present in the modeling. As discussed in Section 3.2.5, the degree of conservatism in the assessment may vary from hazard group to hazard group. Secondly, the uncertainty in those results needs to be addressed. As discussed throughout this document, many sources of uncertainty exist, and some are common to the analysis of all hazard groups whereas others are specific to a hazard group. Identifying the contributors to the results is essential to identifying the sources of uncertainty relevant to the decision.

Therefore, to characterize the results of the risk assessment adequately, analysis of the results is necessary to identify the contributors to the results being used to compare to the acceptance guidelines, first to determine the level of detail or realism, and then to identify the sources of uncertainty. Section 7.3 addresses this guidance in detail. The approach differs depending on whether the hazard group is modeled using a PRA or whether its contribution is addressed by a bounding assessment.

Step 3: Presentation of Results

As noted earlier, the risk analyst is responsible for ensuring that the conclusions of the risk assessment are documented and communicated clearly. An important part of the documentation of the risk assessment is a discussion of the robustness of, or the confidence in, the conclusions drawn from that analysis. Section 7.4 addresses this guidance in detail.

The remaining subsections of this section discuss addressing uncertainty in structures, systems, and components (SSC) categorization (Section 7.5) and using qualitative approaches to compensate for unquantified (and, therefore, unknown) risk contributions (Section 7.6). In some cases, an analyst may decide not to address a potentially significant contributor to risk quantitatively. This decision may be made for one of a number of reasons, including lack of expertise in the methods or resource limitations. However, if the contributor cannot be demonstrated to be an insignificant contributor to the decision, an alternative approach is needed. One possibility, rather than attempting to quantify the risk impact, is to adopt conditions (e.g., limitation of allowed changes or adoption of compensatory measures) on the implementation of the decision to compensate for the uncertainty by essentially arguing that the implementation does not affect the unquantified portion of risk. A qualitative understanding of the uncertainties as they affect the risk insights also can provide input to defining performance measurement strategies or compensatory measures.

In other cases, when the PRA model is being used to assess the risk implications of a change to the plant design or operation, the quantitative characterization of the impact of a change on the PRA results may be challenging. This challenge can occur, for example, when no agreed-upon theoretical or empirical basis exists for representing the impact on the plant of the issue being evaluated. As a result, the effect on risk cannot be quantified without some degree of arbitrariness. An example is the modeling of the effect of a reduction in special treatment on the unreliability of SSCs. In this case, one solution is the adoption of a performance-monitoring strategy to ensure that the degradation in SSC performance is no larger than that assumed in the demonstration that the change in risk is small.

7.2   Step 1: Description of the Risk Assessment

This section provides guidance on the description of the scope and nature of the risk assessment, assuming the results required and the acceptance guidelines to which they are to be compared have been determined for the application being considered. As discussed in Sections 3.1.1 and 3.1.2, this description is generally defined in guidance documents associated with the decision.

7.2.1 Scope of Risk Assessment

The scope of the risk assessment used for the application is defined using the guidance in Section 6.1. For regulatory applications, the scope of risk contributors that needs to be considered includes all hazard groups and all plant operational states relevant to the decision. In some cases, some hazard groups or plant operational states can be qualitatively shown to be clearly irrelevant to assessing the change in risk associated with the decision. For example, for a decision that involves only the at-power operational state, the low power and shutdown operational states need not be addressed. As another example, if the issue is related to response to a steam generator tube rupture (SGTR), no need exists to consider the risk from fires since they cannot cause a SGTR. However, for decisions for which the RG 1.174

acceptance guidelines apply, a detailed quantitative assessment of the baseline core damage frequency (CDF) and large early release frequency (LERF) needs to be performed if the magnitude of the changes in CDF or LERF are not very small (i.e., if they are not in Region III of the quantitative acceptance guidelines of RG 1.174.

7.2.2 The PRA Model

A PRA is the principal tool for risk evaluation. In accordance with the Commission's Phased Approach to PRA Quality [NRC, 2003e], the risk from each significant risk contributor (hazard group and/or plant operating state [POS]) should be addressed using a PRA model that is performed in accordance with a consensus standard for that risk contributor (hazard group and/or POS) that has been endorsed by the staff. A significant risk contributor is one whose consideration in the decision can make a difference to the decision. As discussed in Section 6, contributors shown to be insignificant can be addressed by using conservative assessments or being screened.

The PRA model[17] being used as a basis for the generation of the risk results for significant contributors to support the application is described in terms of its scope (i.e., hazard groups and POSs).

As part of the development of the PRA models for the hazard groups, sources of uncertainty need to be identified and characterized. Guidance for addressing the corresponding supporting requirements of the American Society of Mechanical Engineers/American Nuclear Society (ASME/ANS) PRA Standard [ASME/ANS, 2009] is included in Section 4 and Section 5 for parameter and model uncertainty, respectively, and in Sections 2 and 3 of the Electric Power Research Institute (EPRI) Report 1016737 [EPRI, 2008].

7.2.3 Using the PRA Model

In documenting the risk assessment, the approach used to provide the required results is described and the risk metrics being evaluated are identified. For example, for a RG 1.174 assessment, the metrics are incremental CDF (ΔCDF) and CDF, and incremental LERF (ΔLERF) and LERF.

Also, in documenting the risk assessment, the calculations performed using the PRA model are described. The cause-effect relationship that relates the subject of the application to the parts of the PRA model is identified, and modifications to the PRA model or its constituent parts to evaluate the risk implications are defined. Depending on the application, these modifications can range from changes to parameters in the PRA model, to introduction of new events, to changes in the logic structure.

Several application-specific guidance documents (e.g., RG 1.174, RG 1.177 [NRC, 1998b], Nuclear Energy Institute [NEI] 00-04 [NEI, 2005b]) and the PSA Applications Guide [EPRI, 1995] provide guidance on using a PRA model to address a specific issue. For some applications, only a portion of the complete PRA model is needed. For other applications, such as the identification of the significant contributors to risk, the complete PRA is needed. In addition, the RG 1.174 acceptance guidelines are structured so that even if only a portion of the PRA results are required to assess the change in CDF and LERF (depending on the magnitude

---

[17] In this section, the term PRA model can be interpreted as a fully integrated model in which all hazard groups are combined into one logic structure, the set of individual PRA models for each hazard group, or a mixture.

of that change), an assessment of the base PRA risk metrics (e.g., base CDF and LERF) may be needed.

As discussed in Section 7.3, these results and their associated uncertainties are analyzed in detail to determine the robustness of the results obtained from the PRA model.

7.3 Step 2: Comparison of Risk Assessment Results with Acceptance Guidelines

The comparison with the acceptance guidelines is performed. The primary tool for evaluating the risk contributions is a PRA model. However, the PRA model will generally be supplemented by qualitative or quantitative screening of those hazard groups that are either irrelevant or insignificant. In addition, some contributors to the risk evaluation may be evaluated using bounding or conservative approaches. Generally, the comparison needs to be performed using the sum of the appropriate point estimates, typically the mean value, for each of the risk contributors. For contributors that are not significant to the decision, the point estimates may be conservative estimates. As discussed below, this comparison needs to be done with an understanding of the potential for variability in PRA results driven by the choice of approximations or assumptions made for convenience rather than as a result of model uncertainty.

7.3.1 Level of Detail

To develop confidence in the conclusions of the risk assessment, an analysis is performed to assess the robustness of the comparison with the acceptance guidelines with respect to the sources of epistemic uncertainty, as discussed in Section 7.3.3. However, even in the absence of parameter, model, or completeness uncertainty, uncertainties exist that are a result of the level of detail and approximations used to develop the model. As an example, differences can arise in PRA models performed by two groups of analysts for the same plant because one model may be developed to a lower level of detail than the other. For example, when developing a PRA model for a boiling water reactor (BWR), one analyst may choose to include the fire water system as an additional low-pressure injection system whereas another might not. The latter may have decided that the contribution to the frequency of core damage of the accident sequences resulting from failure of the low-pressure injection function is low enough without taking into account systems in addition to the low-pressure coolant injection and low-pressure core spray.

In addition, differences can occur in the choice of modeling assumptions and approximations made to limit the need for potentially resource-intensive detailed analysis. Examples of such modeling approximations include the subsuming of initiating events into groups, assuming failure of SSCs due to fires when detailed cable routing information is not readily available, and using screening approaches for addressing the less significant contributors. Generally, a less detailed model would be expected to be slightly conservative with respect to a more detailed model. The level of conservatism incorporated in a PRA model needs to be considered when using the results in a risk-informed application, as discussed later in this section. Although this may not matter for many applications, it could be an important factor for some applications.

However, PRA models developed in accordance with the ASME/ANS standard to meet the same capability categories for the same set of supporting requirements would be expected to have identified the same risk-significant sequences. Therefore, the differences between models arising from different levels of detail or different modeling approximations should be in the less significant contributors. Therefore, the differences between the model results that are a consequence of the level of detail would be expected to be minor for the base PRA.

Consequently, in reality, care has to be taken when drawing conclusions by comparing differences in quantitative results from two analyses that purport to represent the same issue but differ by a very small amount. The scale for what this small amount should be is not defined. However, using the acceptance guidelines for RG 1.174 as the basis for generally defining "small," a very small change is defined as being less than 10^{-6} per reactor year (ry) for ΔCDF or 10^{-7}/ry for ΔLERF. These values were chosen recognizing that these are assessments of changes that could be negated by a (legitimate) refinement of the model and thus represent a measure of the limit of discrimination afforded by a PRA model.

It is particularly important to recognize differences in levels of detail or realism with which the different hazard groups and also the different contributors within each hazard group are modeled when those differences can influence the results. Therefore, when interpreting the results of the comparison, it is essential not only to focus on the numerical integrated result but, as discussed in the next section, to also determine the relative realism of the main contributors. If there are contributors for which the evaluation is considered to be particularly conservative (in that conservatism could lead to a challenge in meeting the risk acceptance criteria), the analyst needs to provide a convincing demonstration of the extent of the conservatism. It may be appropriate in this case to refine the PRA model to remove some conservatism.

The identification of the contributors to the results also is needed to identify those sources of epistemic uncertainty that have the potential to influence a decision (i.e., the key sources of uncertainty) (see Section 5). Some of the sources of uncertainty are relevant to several hazard groups (e.g., those that are relevant to the plant response model for a fire or seismic PRA are also relevant to the internal events PRA). Others are unique to the analysis of the specific hazard group.

Finally, consideration also has to be given to the nonquantified biases (e.g., the conservatism resulting from the modeling approach or the improvement to safety from programs or design not captured in the PRA model).

The first step in addressing the robustness of the results, therefore, is to identify the significant contributors to the results and to assess how realistic they are.

7.3.2 Interpretation of the Results of the Risk Assessment

To identify the sources of uncertainty and related assumptions that could have an impact on the results, the significant contributors to the results are identified. In this context, significance should be measured with respect to the sum of the contributions from all hazard groups to the assessed metrics that are important to the decision. The significant contributors can be categorized in different ways. The hierarchical approach proposed here is first to decompose the results into hazard groups, followed by an identification of the significant accident sequences or cutsets, and then by an identification of the significant basic events, as necessary. While identifying the significant contributors, it is necessary to understand how approximations and the level of detail included in the PRA model can bias the determination of significance. The following sections discuss reasons for this decomposition.

Other approaches to decomposing the results are possible. Some PRAs, for example, are fully integrated models that include in a single logic model all the accident sequences caused by the hazard groups modeled. For such models, the contributions from the different hazard groups may be comingled in the logic structure, but it is still possible to identify contributions from the different hazard groups. Moreover, even for these models, the analyses for the different hazard groups are typically performed separately before they are combined into a single model. The concern here is not with the form of the final model but to identify the approximations and level of detail that went into the development of its constituent parts.

7.3.2.1 Decomposition into Hazard Groups

The reason for decomposing by hazard group is the following. PRA methods exist, in principle, for analyzing each of the potentially significant hazard groups (i.e., internal events, internal fires, external events) for the different POSs. The methods employed to address internal fires, external events, and for POSs other than at power typically make use of the models developed for the analysis of internal events at power to address the plant response to the various challenges to the plant. For this reason, the assessments of the risk for these hazard groups are subject to the sources of uncertainty associated with those models. However, in addition, each of the PRA analyses for these other scope items has its own unique sources of uncertainty. For example, a fire PRA has uncertainties associated with the modeling of fire growth and suppression as well as other areas of the analysis, such as the potential for fire damage; a seismic PRA has uncertainties associated with the assessment of the seismic hazard and with the structural response of the SSCs.

Moreover, the models for the different hazard groups are often developed to different levels of detail. In particular, the set of mitigating equipment that is credited with fulfilling the critical safety functions may, for one reason or another, differ from one hazard group to the next. For example, when analyzing internal fires, only those systems that are known to be unaffected by fires might be credited. The demonstration that a particular system that is not part of the safe-shutdown train is not affected by fire may involve cable tracing. If the fire-initiated CDF is low enough without taking credit for that system, it may be decided not to perform cable tracing, and the system would not be included in the fire PRA even if credit were taken for that system in the internal events PRA. Thus, the PRA models for the different hazard groups may reflect a different level of detail in modeling plant response.

In addition, when constructing a PRA model for the risk from a specific hazard group, screening approaches are typically employed to limit the number of scenarios that need to be modeled. The specific approach to screening varies between the PRA models used for the different risk contributors. This can result in some of the models yielding results of differing degrees of realism or conservatism that can be particularly significant when assessing the impact of changes to the plant. In addition, it is claimed that, for some risk contributors, particularly fires and seismic events, the methods used are intrinsically conservative and the uncertainties in the results are of a different nature and magnitude. For the purposes of comparison with acceptance guidelines or criteria, it is necessary to add the results from the various contributors to risk. Thus, the result needs to be interpreted carefully to take into account the differing degrees of realism and potential conservative biases associated with each of the constituent PRA models. This is particularly important when the risk results challenge the acceptance guidelines.

Some hazard groups may be analyzed using bounding or conservative approaches, and these typically will not give information on the significant accident sequences, cutsets, or basic events.

For these contributors, the treatment of uncertainty is relatively straightforward in that the contribution to risk is clearly conservative. However, as discussed later, in some applications a conservative treatment can bias conclusions in a nonconservative manner by lessening the significance of SSCs, for example.

7.3.2.2 Decomposition into Significant Accident Sequences or Cutsets

The identification of the significant accident sequences and also the significant cutsets provides another perspective of the results. A review will identify those sources of uncertainty that are common to all hazard groups as well as those that are unique to hazard groups. As an example, station blackout scenarios may be caused by a number of different hazard groups and, therefore, any uncertainties associated with the plant response (e.g., the reactor coolant pump seal loss-of-coolant-accident model) would be common to all hazard groups.

7.3.2.3 Decomposition into Significant Basic Events

The identification of significant basic events is performed using importance analyses. Because importance measures such as the Fussell-Vesely Importance or Risk Achievement Worth are dependent on the total model, differences in the approaches used for different hazard groups can affect the measures significantly. Therefore, it is advisable that the importance measures be evaluated for each hazard group separately as well as integrally. If the importance of a component is significantly different for one hazard group from that for another, the reason should be identified. A low overall Fussell-Vesely importance could result from the fact that the component is not a contributor to the mitigating strategy for a specific hazard group. On the other hand it could be because the model did not take credit for that component. In the latter case, the overall importance of the component would be artificially low, and this should be recognized when using the results. On the other hand, not taking credit for a system in the PRA for one or more hazard groups would artificially raise the Risk Achievement Worth of those SSCs for which the uncredited system provides an alternate means of achieving their function.

7.3.3 Assessment of Impact of Uncertainty on the Comparison with Acceptance Guidelines

As discussed in Section 3, epistemic uncertainty is generally categorized into three types: parameter, model, and completeness uncertainty. Because they are characterized in different ways, the approaches to addressing them are different as discussed below. The analysis of uncertainty is best carried out sequentially, dealing first with parameter uncertainty and then addressing model uncertainty. This needs to be done both for the hazard groups and POSs separately and in combination.

7.3.3.1 Parameter Uncertainty

Section 4 describes methods for addressing parameter uncertainty in the calculation of PRA results. Because the impact of parameter uncertainty can be addressed in terms of a probability distribution on the numerical results of the PRA, it is straightforward to compare a point value, be it the mean, the 95th percentile, or some other representative value with an acceptance guideline or criterion that is intended for use with that type of result.

The representative value to be used should be that specified when the acceptance guideline or criterion is established. For most regulatory applications, that value is specified to be the mean value when integral metrics such as CDF, LERF, incremental conditional core damage probability, etc. are being calculated. The rationale for using the mean value is provided in SECY-97-221 [NRC, 1997a] and SECY-97-278 [NRC, 1997b]. The mean values referred to are the arithmetic means of the probability distributions that result from the propagation of the uncertainties on the input parameters.

Although a formal propagation of the uncertainty is the best way to correctly account for epistemic uncertainties that arise from the use of the same parameter values for several basic event probability models, under certain circumstances, a formal propagation of uncertainty may not be necessary if it can be demonstrated that the epistemic correlation (also called state-of-knowledge correlation) is unimportant. This will involve, for example, a demonstration that the bulk of the contributing scenarios (cutsets or accident sequences) do not involve multiple events that rely on the same parameter for their quantification. Section 4.3 discusses this, and additional guidance is given in EPRI 1016737.

It should be noted that if the PRA model has explicitly represented model uncertainties by, for example, using a discrete probability distribution over the set of reasonable alternate assumptions, and propagated them through the analysis (see discussion below), this also would be factored into the mean value.

When the metrics are importance measures, the computer codes used for their evaluation typically rely on the use of point estimates rather than the complete uncertainty distribution. However, it should be noted that the values of the basic event probabilities in the calculation should represent the means of the probability distributions on the basic event probabilities. Ideally, a formal propagation of parameter uncertainty through the PRA model also would be carried out to obtain the mean values of the importance measures. Section 4.4 further discusses treating parameter uncertainty in importance measures.

7.3.3.2 Model Uncertainty

The analysis of the significant contributors results in an identification of the subset of the relevant sources of model uncertainty that could have an impact on the results. The second level of analysis addresses these model uncertainties. In the context of decisionmaking, it is necessary to assess whether these uncertainties have the possibility of changing the evaluation of risk significantly enough to alter a decision. Section 5 discusses methods for identifying and assessing the impact of key sources of model uncertainty. As discussed in Section 5, this assessment can take the form of reasonable and well-formulated sensitivity studies or qualitative arguments.

In this context, "reasonable" is interpreted as implying that some precedent exists for the alternative assumption or model represented by the sensitivity study(ies), such as previous use by other analysts, and that a physically reasonable and documented basis exists for the alternative. It is not necessary for the search for alternatives to be exhaustive and arbitrary. The sensitivity studies provide a discrete set of results that demonstrate the impact of the alternative models or assumptions on the mean values calculated as discussed above.

In principle, it is possible to associate a probability with each of the sensitivity cases to express the analysts' relative degrees of belief in the members of the set of assumptions. The resulting probability distribution could be used to provide a mean value that, as indicated in Section 7.3.3.1 above, could be used for comparison with the acceptance guidelines. Such a formal approach to the representation of uncertainty has been adopted in a number of studies that involved expert elicitation to address complex issues with significant model uncertainty. Examples include NUREG 1150 [NRC, 1990] and the guidelines for seismic hazard evaluation provided by the Senior Seismic Hazard Analysis Committee (SSHAC) [LLNL, 1997]. Although this approach to displaying uncertainty is theoretically appealing, for most of the issues that are expected to be encountered it is considered to be more useful to present the results of the sensitivity studies separately. However, for this to be meaningful to a decisionmaker it is incumbent on the analyst to provide an assessment of the credibility of each of the separate assumptions, particularly when the alternatives would lead to different recommendations, as discussed further below.

As discussed in Section 5, for some sources of model uncertainty, the sensitivity analyses are performed by changing the values of some of the input parameters. For others, changes in the logic structure may be required. Two potentially significant cross-cutting sources of uncertainty are Human Reliability Analysis and Common Cause Failure Analysis. Approaches to addressing these sources of uncertainty are discussed below.

Human Reliability Analysis: If the human failure events (HFEs) have been identified and defined by following good practices [NRC, 2005] consistent with the requirements of the ASME/ANS standard, one of the most significant sources of uncertainty is in the choice of the quantification tool (human reliability analysis [HRA] method) for estimating human error probabilities (HEPs). Although it would in principle be feasible to perform the HRA using an alternate method, this might not be practical due to limited expertise with using the alternate method. An acceptable approach is to perform a sensitivity study varying all the HEPs by the same factor. The magnitude of the factor should be chosen taking into account a number of issues, including the uncertainty range dictated by the HRA method, but also a comparison of the HEPs derived for similar HFEs in different PRAs using different HRA methods. This should be done both for an increase in the HEPs and by a decrease for the following reasons. An optimistic evaluation of the HEPs can lead to the lessening of the importance of the SSCs that appear in the same cutsets or accident sequences as the corresponding HFEs. On the other hand, a conservative evaluation can lead to masking the importance of other contributors, particularly those in cutsets and sequences not involving the HFEs.

Common Cause Failure (CCF) Analysis: A major uncertainty with respect to common cause failure analysis is related to the estimation of the parameters of the models used. There is general agreement that the multi-parameter models, such as the alpha model of the multiple Greek letter method, are the appropriate models to use to represent the CCF terms in the fault trees. If applied, using the same interpretation of the data, these models will give consistent results. However, considerable uncertainty exists in the interpretation and manipulation of the data to derive the parameters of the models. An acceptable approach to addressing this is to perform a set of sensitivity analyses, similar to those performed for the HFEs, in which the CCF parameters are changed by a common factor. The reasoning is the same as for the HFEs; namely, the need to make sure that the CCF analyses neither artificially increase nor decrease the importance of related SSCs.

When the results of the sensitivity studies confirm that the guidelines are still met even under the alternative assumptions (i.e., conclusion generally remains the same with respect to the acceptance guideline), the principle of risk-informed regulation dealing with acceptable risk impact can be determined to be met with confidence.

However, when some alternative hypotheses lead to a significant change in the relationship of the PRA result to the acceptance criteria, the analyst should provide the decisionmaker with the basis for considering the credibility of the alternatives. For example, if reasons can be given as to why they are not appropriate for the particular application or for the particular plant, they need not be taken into account. If, however, the decisionmaker does not have sufficient confidence that the alternative hypothesis that challenges the decision criteria can be discounted, the tendency will be to err on the side of caution. Alternatively, the analysis can be used to identify candidates for compensatory actions or increased monitoring (see Section 7.5).

As discussed in Section 5, it is necessary not only to consider the impact of sources of uncertainty individually, but also in logical combinations. Section 5 provides guidance on the identification of logical combinations.

When the risk estimate is derived from a number of PRA models providing estimates for the contributing hazard groups and if the model uncertainty (or uncertainties if considered as a logical group) affects several of the constituent PRA models, the analysis of model uncertainty should be performed in an integral manner. In other words, the sensitivity analysis of a model uncertainty that is pertinent to the plant response model should be demonstrated for the sum of all the affected risk contributors.

Sources of uncertainty that affect the significant contributors (at the level of hazard groups and POSs, accident sequences, cutsets, and SSCs) to the relevant results also should be considered together. Clearly, this could become intractable if some degree of reasonableness is not adopted. It is anticipated that the number of sources of model uncertainty that can affect the significant contributors sufficiently to alter the decision will in general be limited, and it is on these that attention should be focused.

7.3.3.3 Completeness Uncertainty

When a particular risk contributor is not evaluated by a PRA model, then either the effect on the application has to be bounded and shown not to be significant or other measures have to be taken to ensure that the assumption of no risk increase is supported (see Section 7.5). In this section, what is classified as a completeness issue is a concern about the limitation of the scope of the model. The issue of completeness of the scope of a PRA can be addressed quantitatively for those scope items for which limited methods are available in principle and, therefore, some understanding of the contribution to risk exists.

For example, the out-of-scope items can be addressed by supplementing the analysis with additional analysis by alternative methods to a fully developed PRA. Section 6 discusses approaches to supplementing the PRA model by using alternate models. These methods can either be approximations or bounding analyses.

However, neither an approximate method nor a bounding analysis can be used in the same way as a full PRA model to fully understand the contributions to risk and thereby gain robust risk insights. Thus, their usefulness is somewhat limited, particularly in applications relying on relative importance measures. The principal use of such methods is to demonstrate that the risk contribution from that contributor, and any change to that risk contribution resulting from a change to the plant, are small. For this to be the case, the result of the analyses should be demonstrably conservative. When this is the case, no need exists to explicitly capture the uncertainty in the risk assessment because it has been demonstrated that the contribution(s) so dispositioned is not significant to the decision.

Care should be taken to make sure that any assumptions made to screen out or bound a hazard group are not invalidated for a specific application. For example, the assumption that tornados can be screened based on the assumption of the existence of tornado missile barriers may be invalid for situations where a barrier is temporarily moved for a particular plant evolution.

The degree to which supplementary arguments can be used to support the claim that the uncertainties do not impact the decision depends on the proximity to the guidelines. When the contributions from the modeled contributors are close to the guideline boundaries, the argument that the contribution from the missing items is not significant needs to be more convincing than when the results are further away from the boundaries and, in some cases, may require additional PRA analyses. When the margin is significant, a qualitative argument may be sufficient.

7.4   Step 3: Presenting the Results of Risk Assessment

The simple presentation of the risk assessment results as a numerical comparison with the acceptance guidelines is not sufficient and should be avoided. The comparison of the appropriate representative value derived from the probability distribution on the corresponding PRA result, be it the mean value or a percentile of the distribution, addresses only the parametric uncertainty analysis, as discussed in Section 4. In RG 1.174, for example, the interpretation of the acceptance guidelines is that they should not be regarded as overly prescriptive. This interpretation is a reflection of the realization that some residual uncertainties are not captured in the mean value. For example, there could be conservatism in the model that would increase the calculated risk metric or things could have been left out that would inappropriately lead to an optimistic assessment of the calculated risk metric. Therefore, acceptance or rejection should not be based on a strict interpretation whereby a numerical result just on one side of the acceptance guideline would be acceptable while one just on the other side would not.

When making a recommendation to a decisionmaker on the basis of the risk assessment results, the analyst should carefully analyze the reasons for acceptance or rejection based on an understanding of the contributors to the results and should take into account the model uncertainties using the techniques of Section 5. Moreover, the analyst should have confidence that the model has sufficient scope to support the conclusions of the analysis using the guidance in Section 6.

When the results from different hazard groups need to be combined for comparison with acceptance guidelines, it is important to determine the relative contributions from each of the hazard groups. This is of particular concern because of the potential for differing levels of conservatism associated with the modeling of different hazard groups, as discussed in Section 7.3.1. This becomes most important when the total risk metric value approaches the boundary between regions of acceptability. When the representative value is well away from the guideline boundaries, any concerns about the impact of a potential bias caused by the choice of level of detail tends to be a lesser concern. However, when the calculated result is above but close to the boundary, then the recognition of the potential conservatism resulting from the chosen level of detail could make the difference between arguing for acceptance or not. The differing level of detail in constituent analyses also is important when considering risk ranking or categorization of SSCs because the absence of SSCs from some of the models will distort the overall results.

Therefore, the presentation of the conclusions to the decisionmaker to support a recommendation on whether the acceptance guidelines have or have not been met should include the following:

- A brief description of the risk assessment, including the scope of the PRA models utilized.

- A qualitative statement of confidence in the conclusion and how it has been reached.

To support the statement of confidence, the analyst should identify the key sources of uncertainty that were addressed. In the event that the analysis results, including all sensitivity analyses, indicate the guidelines are met, the documentation above should be sufficient.

In the event that the base results or a sensitivity case exceeds the acceptance guidelines but acceptance is recommended, the documentation should include, as necessary:

- An identification of any significant conservatism in the modeling.

- A justification of any compensatory measures proposed to compensate for conservatisms in the model.

- A description of why any limitations of applicability are proposed.

- A description of the purpose of a proposed performance monitoring program.

- An assessment of the confidence in the recommendation.

When one or more of the sensitivity results demonstrate that the acceptance guidelines are not met and the recommendation is for rejection of the proposed change, the analyst should provide the decisionmaker with a clear assessment of the credibility of the sensitivity study as a reasonable alternative to the proposed base case analysis.

In all cases, the analysts should maintain as archival material the analyses that support the conclusions, including the analysis of the results, identification of sources of uncertainty, descriptions, and results of sensitivity analyses. The archival documentation of the risk assessment should include the following:

- Identification of the acceptance guidelines used.

- Description of the scope of the base PRA model (hazard groups and plant operating states).
    - Identification of relevant hazard groups screened out of the evaluation and the basis for screening.
    - Identification of hazard groups included in the quantification using conservative analyses.

- Description of how the base PRA model is modified/manipulated for the application (i.e., cause-effect relationship).

- Evaluation of the metrics addressing parameter uncertainty.

- Characterization of the relevant sources of model uncertainty and related assumptions.

- Identification of the key sources of uncertainty and their impact on the results.

7.5 Addressing Uncertainty in SSC Categorization

Categorization of SSCs according to their safety significance is an integral part of many applications, such as the implementation Part 50.69 of Title 10 of the Code of Federal Regulations (10 CFR §50.69). NEI 00-04 [NEI, 2005b] addresses uncertainty primarily by using the point estimates of the importance evaluations to perform the initial categorization and then performing sensitivity studies to vary some of the key groups of parameters (e.g., HEPs, CCF probabilities) and to address some of the model uncertainties identified by the peer review as potentially being significant. These sensitivity studies are used to identify changes in the categorization, and the most conservative categorization is used.

The reasonableness of this approach is justified in two ways. First, EPRI TR 1008905, "Parametric Uncertainty Impacts on Option 2 Safety Significance Categorization" [EPRI, 2003], demonstrates that the impact of performing a full uncertainty analysis to categorize SSCs does not significantly change the categorization resulting from using mean value point estimates in the importance calculations. Second, in regulatory applications, it is necessary to demonstrate that the impact of the change resulting from using the categorization to change plant practices results in small changes to risk using the RG 1.174 acceptance guidelines discussed above.

As discussed in Section 7.3.1.3, the differing level of detail in the models for the different hazard groups requires that care be taken in interpreting the results of using those models to categorize the SSCs. NEI 00-04 adopts an approach that has been found acceptable by NRC.

7.6 Compensating for Unquantified Risk Contributors

When the quantitative analysis of risk does not address all significant risk contributors, or when the impact of a change on risk has not been calculated, it may be acceptable to adopt one of a number of strategies to deal with the unknown (by reason of its being unquantified) impact. These strategies, discussed below, include the following:

- Adopting performance monitoring requirements.
- Limiting the scope of application of plant changes.
- Establishing compensatory measures.

7.6.1 Performance Monitoring Requirements

Monitoring can be used to demonstrate that, following a change to plant design or operational practices, no degradation exists in specified aspects of plant performance. This monitoring is an effective strategy when no predictive model has been developed for plant performance in response to a change. One example of such an instance is the impact of the relaxation of special treatment requirements (in accordance with 10 CFR §50.69) on equipment unreliability. No agreed-upon approach to model this cause-effect relationship has been developed.

Therefore, the approach adopted in NEI 00-04 as endorsed in RG 1.201 [NRC, 2006a] is to assume a multiplicative factor on the SSC unreliability factor that represents the effect and to demonstrate that this degradation in unreliability would have a small impact on risk. Following implementation, a monitoring program has to be established to demonstrate that the assumed factor of degradation is not exceeded. For monitoring to be effective, the plant performance needs to be measurable in a quantitative way, and the criteria used to assess acceptability of performance needs to be realistically achievable given the expected quantity of data that would be expected.

7.6.2 Limiting Scope of Plant Modification

When a PRA model is incomplete in its coverage of significant risk contributors, the scope of implementation of a risk-informed change can be restricted to what the risk assessment can support. For example, if the PRA model does not address fires, the change to the plant could be limited such that any SSCs that would be expected to be used to mitigate the risk from fires would be unaffected. In this way, the contribution to risk from internal fires would be unchanged. This is the strategy adopted in NEI 00-04 for categorizing SSCs according to their risk significance when the PRA being used does not address certain hazard groups.

7.6.3 Use of Compensatory Measures

The purpose of implementing compensatory measures is to neutralize the expected negative impact of some feature of plant design or operation on risk. For example, a fire watch may be established to compensate for a faulty fire barrier or the temporary removal of a barrier. Another example is the implementation of a manual action (suitably proceduralized and accounted for in training) to replace an automatic actuation of a system (e.g., the initiation of depressurization of a BWR following loss of high-pressure injection that is necessitated by the inhibition of the automatic depressurization system function). In the latter case, this is modeled in the PRA. However, if the compensatory action is not modeled in the PRA but is being proposed as a means to provide an argument of why the PRA result is conservative, understanding the scenarios for which the compensatory measures are designed is necessary to provide confidence in their value. For example, establishing that the fires that the fire watch is intended to mitigate are slow growing fires adds confidence to its value. On the other hand, if a high-energy arcing fault were a significant contributor to the fires in an area, a fire watch would be ineffective.

8.0 REFERENCES

[Apost., 1981]   Apostolakis, G., and Kaplan, S., "Pitfalls in Risk Calculations," *Reliability Engineering,* Vol. 2, pp. 135-145, 1981.

[Apostolakis, 1994]   Apostolakis, G., "A Commentary on Model Uncertainty", in Proceedings of Workshop 1 in Advanced Topics in Risk and Reliability Analysis, Model Uncertainty, its Characterization and Quantification, NUREG/CP-0138, October, 1994.

[ASME/ANS, 2009]   ASME/American Nuclear Society, "Standard for Level 1/Large Early Release Frequency Probabilistic Risk Assessment for Nuclear Power Plant Applications," ASME/ANS RA-Sa-2009, March 2009.

[ACRS, 2003a]   Advisory Committee on Reactor Safeguards, Letter from M. Bonaca, ACRS Chairman, to Chairman Diaz, NRC, "Proposed Resolution of Public Comments on Draft Regulatory Guide (DG)-1122, 'An Approach for Determining the Technical Adequacy of Probabilistic Risk Results for Risk-Informed Decision Making,'" Washington, D.C., April 21, 2003.

[ACRS, 2003b]   Advisory Committee on Reactor Safeguards, Letter from M. Bonaca, ACRS Chairman, to Chairman Diaz, NRC, "Improvement of the Quality of Risk Information for Regulatory Decision Making," Washington, D.C., May 16, 2003.

[EPRI, 1985]   Electric Power Research Institute, "Classification and Analysis of Reactor Operating Experience Involving Dependent Events," EPRI NP-3967, Palo Alto, CA, June 1985.

[EPRI, 1988]   Electric Power Research Institute, "Procedures for Treating Common Cause Failures in Safety and Reliability Studies," NUREG/CR-4780, EPRI NP-5613, PLG-0547, Vol. 1, Palo Alto, CA, January 1988.

[EPRI, 1995]   Electric Power Research Institute, "Probabilistic Safety Assessments Applications Guide," EPRI TR 105396, Palo Alto, CA, August 1995.

[EPRI, 2003]   Electric Power Research Institute, "Parametric Uncertainty Impacts on Option 2 Safety Significance Categorization," EPRI TR 1008905, Palo Alto, CA, June 2003.

[EPRI, 2004]   Electric Power Research Institute, "Guideline for the Treatment of Uncertainty in Risk-Informed Applications: Technical Basis Document," EPRI TR 1009652, Palo Alto, CA, December 2004.

[EPRI, 2005]   Electric Power Research Institute, "EPRI/NRC-RES Fire PRA Methodology for Nuclear Power Facilities," NUREG/CR-6850.

[EPRI, 2006]   Electric Power Research Institute, "Guideline for the Treatment of Uncertainty in Risk-Informed Applications: Applications Guide," EPRI 1013491, Palo Alto, CA, 2006.

[EPRI, 2008]   Electric Power Research Institute, "Treatment of Parameter and Model Uncertainty for Probabilistic Risk Assessments," EPRI TR 1016737, Palo Alto, CA, December 2008.

[Helton, 1996]   Helton, J.C., and Burmeister, D.E., (editors), "Treatment of Aleatory and Epistemic Uncertainty,", special issue of *Reliability Engineering and System Safety,* Vol. 54, 2-3 (1996)

[INL, 1998a]   Idaho National Engineering and Environmental Laboratory, "Common-Cause Parameter Estimations," NUREG/CR-5497, INEEL/EXT-97-01328, Idaho Falls, ID, October 1998.

[INL, 1998b]   Idaho National Engineering and Environmental Laboratory, "Common-Cause Failure Database and Analysis System: Overview," NUREG/CR-6268, INEEL/EXT-97-00696, Idaho Falls, ID, June 1998.

[INL, 1998c]   Idaho National Engineering and Environmental Laboratory, "Guidelines on Modeling Common-Cause Failures in Probabilistic Risk Assessments," NUREG/CR-5485, INEEL/EXT-97-01327, Idaho Falls, ID, November 1998.

[LLNL, 1997]   Senior Seismic Hazard Analysis Committee (SSHAC), "Recommendations for Probabilistic Seismic Hazard Analysis: Guidance on Uncertainty and Use of Experts," UCRL-ID-122160, NUREG/CR-6372, Lawrence Livermore National Laboratory, April 1997.

[NEI, 2005a]   "Process for Performing Follow-on PRA Peer Reviews Using the ASME PRA Standard," NEI-05-04, Washington, D.C., January 2005.

[NEI, 2005b]   Nuclear Energy Institute, "10 CFR 50.69 SSC Categorization Guideline", NEI-00-04, ADAMS Accession #ML052910035, Washington, D.C., July 2005.

[NEI, 2006]   Nuclear Energy Institute, "Probabilistic Risk Assessment (PRA) Peer Review Process Guidance," Update to Revision 1, NEI-00-02, Washington, D.C., October 2006.

[NRC, 1975]   U.S. Nuclear Regulatory Commission, "Standard Review Plan for the Review of Safety Analysis Reports for Nuclear Power Plants," NUREG-0800, Washington, D.C., November 1975.

[NRC, 1986]   U.S. Nuclear Regulatory Commission, "Safety Goals for the Operations of Nuclear Power Plants: Policy Statement," *Federal Register*, Vol. 51, p. 30028 (51 FR 30028), Washington, D.C., August 4, 1986.

[NRC, 1990]   U.S. Nuclear Regulatory Commission, "Severe Accident Risk: An Assessment for Five U.S. Nuclear Power Plants," NUREG-1150, Washington, DC, December 1990.

[NRC, 1991]   U.S. Nuclear Regulatory Commission, NUREG-1407, "Procedural and Submittal Guidance for the Individual Plant Examination of External Events (IPEEE) for Severe Accident Vulnerabilities," Washington, D.C., June 1991.

[NRC, 1995]   U.S. Nuclear Regulatory Commission, "Final Policy Statement 'Use of Probabilistic Risk Assessment (PRA) Methods in Nuclear Regulatory Activities'," Washington, D.C., 1995.

[NRC, 1996]   U.S. Nuclear Regulatory Commission, "Branch Technical Position on the Use of Expert Elicitation in the High-Level Radioactive Waste Program," NUREG-1563, Washington, DC, November 1996.

[NRC, 1997a]   U.S. Nuclear Regulatory Commission, "Acceptance Guidelines and Consensus Standards for Use in Risk-Informed Regulation," SECY-97-221, Washington, D.C., September 30, 1997.

[NRC, 1997b]   U.S. Nuclear Regulatory Commission, "Final Regulatory Guidance on Risk-Informed Regulation: Policy Issues," SECY-97-287, Washington, D.C., December 12, 1997.

[NRC, 1998a]   U.S. Nuclear Regulatory Commission, "An Approach for Plant-Specific, Risk-Informed Decision Making: Inservice Testing," RG 1.175, Washington, D.C., August 1998.

[NRC, 1998b]   U.S. Nuclear Regulatory Commission, "An Approach for Plant-Specific, Risk-Informed Decision Making: Technical Specifications," RG 1.177, Washington, D.C., August 1998.

[NRC, 1999]   U.S. Nuclear Regulatory Commission, "Commission Issuance of White Paper on Risk-informed and Performance-based Regulation," Yellow Announcement # 019, Washington, D.C., dated March 11, 1999.

[NRC, 2002]   U.S. Nuclear Regulatory Commission, "An Approach for Using Probabilistic Risk Assessment in Risk-Informed Decisions on Plant-Specific Changes to the Licensing Basis," RG 1.174, Washington, D.C., November 2002.

[NRC, 2003a]   U.S. Nuclear Regulatory Commission, Letter from William Travers, Executive Director for Operations, to M. Bonaca, ACRS Chairman, "Proposed Resolution of Public Comments on Draft Regulatory Guide (DG) 1122, 'An Approach for Determining the Technical Adequacy of Probabilistic Risk Results for Risk-Informed Decision Making'," Washington, D.C., June 4, 2003.

[NRC, 2003b]   U.S. Nuclear Regulatory Commission, Letter from William Travers, Executive Director for Operations, to M. Bonaca, ACRS Chairman, "Improvement of the Quality of Risk Information for Regulatory Decision Making," Washington, D.C., August 4, 2003.

[NRC, 2003c]   U.S. Nuclear Regulatory Commission, "An Approach for Plant-Specific Risk-Informed Decision Making for Inservice Inspection of Piping," RG 1.178, Washington, D.C., September 2003.

[NRC, 2003d]   U.S. Nuclear Regulatory Commission, "Safety Evaluation of Topical Report WCAP-15603, Revision 1, 'WOG 2000 Reactor Coolant Pump Seal Leakage Model for Westinghouse PWRs," ADAMS Accession No. ML031400376, Washington, D.C., May 30, 2003.

[NRC, 2003e]   U.S. Nuclear Regulatory Commission, "PRA Quality Expectations and Requirements," Staff Requirements Memorandum COMNJD-03-0002, Washington D.C., December 18, 2003.

[NRC, 2004]   U.S. Nuclear Regulatory Commission, "Regulatory Analysis Guidelines of the U.S. Nuclear Regulatory Commission (rev. 4)", NUREG/BR-0058, Washington, D.C., September 2004.

[NRC, 2005]   U.S. Nuclear Regulatory Commission, "Good Practices for Implementing Human Reliability Analysis," NUREG-1792, Washington, D.C., April 2005.

[NRC, 2006a]   U.S. Nuclear Regulatory Commission, "Guidelines for Categorizing Structures, Systems, and Components in Nuclear Power Plants According to Their Safety Significance," RG 1.201, Washington, D.C., May 2006.

[NRC, 2006b]   U.S. Nuclear Regulatory Commission, "Risk-Informed, Performance-Based Fire Protection for Existing Light-Water Nuclear Power Plants," RG 1.205, Washington, D.C., May 2006.

[NRC, 2007a]   U.S. Nuclear Regulatory Commission, "An Approach for Determining the Technical Adequacy of Probabilistic Risk Assessment Results for Risk-Informed Activities," RG 1.200, Revision 1, Washington, D.C., January 2007.

[NRC, 2007b]   U.S. Nuclear Regulatory Commission, "Combined License Applications for Nuclear Power Plants (LWR Edition)" RG 1.206, Washington, D.C., June 2007.

[NRC, 2007c]   U.S. Nuclear Regulatory Commission, "Standard Review Plan, Section 19.0, 'Probabilistic Risk Assessment and Severe Accident Evaluation for New Reactors,'" Revision 2, Washington, D.C., June 2007.

[Reinert, 2006] Reinert, J. M., and Apostolakis, G. E., "Including Model Uncertainty in Risk-informed Decision Making," Annals of Nuclear Energy, Vol. 33, pp. 354-369, January 19, 2006.

[SNL, 1988]   Sandia National Laboratories, "Approaches to Uncertainty Analysis in Probabilistic Risk Assessment," SAND87-0871, NUREG/CR-4836, Albuquerque, NM, January 1988.

[SNL, 1989]   Sandia National Laboratories, "Analysis of Core Damage Frequency from Internal Events: Expert Judgment Elicitation," SAND86-2084, NUREG/CR-4550, Vol. 2, April 1989.

[SNL, 1992a]   Sandia National Laboratories, "Analysis of the LaSalle Unit 2 Nuclear Power Plant: Risk Methods Integration and Evaluation Program (RMIEP)," SAND92-0537, NUREG/CR-4832, Albuquerque, NM, October 1992.

[SNL, 1992b]   Sandia National Laboratories, "Methods for External Event Screening Quantification: Risk Methods Integration Evaluation Program (RMIEP) Methods Development," SAND87-7156, NUREG/CR-4839, Albuquerque, NM, July 1992.

[SNL, 2003]   Sandia National Laboratories, "Handbook of Parameter Estimation for Probabilistic Risk Assessment," SAND2003-3348P, NUREG/CR-6823, Albuquerque, NM, September 2003.

[Zio, 1996]   E. Zio and G. Apostolakis, "Two Methods for the Structured Assessment of Model Uncertainty by Experts in Performance Assessments of Radioactive Waste Repositories," Reliability Engineering and System Safety, Vol. 54, no. 2-3, pp. 225–241, December 1996.